The Rhetoric of Science

The Rhetoric of Science

Alan G. Gross

Harvard University Press

Cambridge, Massachusetts

London, England   1990

To Suzanne

Library of Congress Cataloging-in-Publication Data
Gross, Alan G.
The rhetoric of science / Alan G. Gross.
p.   cm.
Includes bibliographical references.
ISBN 0-674-76873-6 (alk. paper)
1. Communication in science—Philosophy. 2. Rhetoric—Philosophy.
I. Title.
Q223.G77   1990
501'.4—dc20
90-4006
CIP

Acknowledgments

For their valuable comments on various chapters, I should like to thank Rod Bertolet, Arthur Fine, A. T. Grafton, David Hull, Tony Lamb, Thomas McCarthy, Robert K. Merton, Doug Mitchell, Victor Namias, George Sefler, Herb Simons, and Charles Tseng. Arthur Fine, Joe Williams, and my two editors, Gunder Hefta and Howard Boyer, offered encouragement at crucial moments. The University of Chicago, and especially the Regenstein and Crerar libraries, extended a hospitality without which this work would have been impossible. For financial support, I should like to thank the Lilly Foundation, Purdue University Calumet, and the National Endowment for the Humanities; without the Endowment's Summer Seminars this project would have been neither begun nor completed.

To Gloria and Harold Fromm, I owe more than thanks. Their encouragement has sustained me through a decade of work, and Harold's ruthless criticism has improved the style and soundness of every chapter.

Earlier versions of several chapters first appeared in the following publications:

Chapters 2, 4, 8, and 12 originally appeared in various publications of the Speech Communication Association.

Chapter 3 was originally published in *The Rhetorical Turn: Invention and Persuasion in the Conduct of Inquiry*, ed. Herbert Simons (Chicago: University of Chicago Press, 1990). © 1990 by The University of Chicago. All rights reserved.

Chapter 6 was originally published in *Journal of Technical Writing and Communication* 15 (1985): 15–26. Reprinted by permission of Baywood Publishing Company.

Chapter 7 originally appeared (as a precis) in *Argumentation: Perspectives and Approaches. Proceedings of the First International Conference on Argumentation*, vol. 3A (Dordrecht: Foris, 1987), pp. 347–356. Reprinted by permission of Foris Publications.

Chapter 11 originally appeared in *Rhetoric in the Human Sciences*, ed. Herbert W. Simons (London: Sage, 1989), pp. 89–108. Reprinted by permission of Sage Publications.

Contents

PART I

The Relation of Rhetoric to Science

CHAPTER 1

Rhetorical Analysis

We readily concede that the law courts and the political forum are special cases of our everyday world, a world in which social reality is uncontroversially the product of persuasion. Many of us can also entertain a possibility Aristotle could never countenance: the possibility that the claims of science are solely the products of persuasion. We live in an intellectual climate in which the reality of quarks or gravitational lenses is arguably a matter of persuasion; such a climate is a natural environment for the revival of a rhetoric that has as its field of analysis the claims to knowledge that science makes.

Rhetorically, the creation of knowledge is a task beginning with self-persuasion and ending with the persuasion of others. This attitude toward knowledge stems from the first Sophistic, an early philosophical relativism made notorious by Socrates. In spirit, the *Rhetoric,* my master theoretical text, is also Sophistic, its goal "to find out in each case the existing means of persuasion." It is a spirit, however, that Aristotle holds firmly in check by limiting the scope of rhetoric to those forums in which knowledge is unquestionably a matter of persuasion: the political and the judicial. If scientific texts are to be analyzed rhetorically, this Aristotelian limitation must be removed; the spirit of the first Sophistic must roam free.

Whether, after rhetorical analysis is completed, there will be left in scientific texts any constraints not the result of prior persuasion, any "natural" constraints, remains for the moment an open question. In the meantime, as rhetorical analysis proceeds unabated, science may be progressively revealed not as the privileged route to certain knowledge but as another intellectual enterprise, an activity that takes its place beside, but not above, philosophy, literary criticism, history, and rhetoric itself.

The rhetorical view of science does not deny "the brute facts of nature"; it merely affirms that these "facts," whatever they are, are not science itself, knowledge itself. Scientific knowledge consists of the current answers to three questions, answers that are the product of professional conversation: What range of "brute facts" is worth investigating? How is this range to be investigated? What do the results of these investigations mean? Whatever they are, the "brute facts" themselves mean nothing; only statements have meaning, and of the truth of statements we must be persuaded. These processes, by which problems are chosen and results interpreted, are essentially rhetorical: only through persuasion are importance and meaning established. As rhetoricians, we study the world as meant by science.

Thirty years ago the humanistic disciplines were more easily definable: historians of science shaped primary sources into chronological patterns of events; philosophers of science analyzed scientific theories as systems of propositions; sociologists of science scrutinized statements aimed at group influence (Markus 1987, p. 43). In the last two decades, however, the humanities have been subject to what Clifford Geertz has called "a blurring of genres." As a result, "the lines grouping scholars together into intellectual communities . . . are these days running at some highly eccentric angles" (1983, pp. 23–24).

David Kohn, Sandra Herbert, and Gillian Beer on Darwin: are they writing intellectual history or literary criticism? Ian Hacking on gravitational lensing: is he doing philosophy or sociology? Arthur Fine on Einstein: is he producing philosophy or intellectual history? Are Steve Woolgar and Karin Knorr-Cetina studying the scientific paper from the point of view of sociology or rhetorical criticism? Is Evelyn Keller's work on Bacon epistemology, psychology, or literary criticism? When Michael Lynch analyzes laboratory shop talk, is he doing ethnomethodology or rhetoric of science?

These intellectual enterprises share a single methodological presupposition; all, to paraphrase Barthes, "star" their texts; all assume with Geertz that "the road to discovering . . . lies less through postulating forces and measuring them than through noting expressions and inspecting them" (1983, p. 34). To address Einstein's philosophy, Fine becomes a historian. Latour and Woolgar discover the intellectual structure of science not through philosophical analysis but through the ethnomethodology of the laboratory. Keller approaches

Bacon's epistemology not by reconstructing his arguments but by analyzing his metaphors; Beer treats the *Origin* less like an argument than like a novel by George Eliot or Thomas Hardy.

Rhetorical analysis describes what all of these scholars of science are doing; it defines the intellectual enterprise of workers as different in outlook and training as Gillian Beer and Steve Woolgar.[1] For such scholars, the speculative knowledge of the sciences is a form of practical knowledge, a vehicle of practical reasoning, whose mark "is that the thing wanted is *at a distance* from the immediate action, and the immediate action is calculated as the way of getting or doing or securing the thing wanted" (Anscombe 1957, p. 79). The *Origin of Species* is speculative knowledge, certainly; from a rhetorical point of view, however, it is also practical knowledge, the vehicle by means of which Darwin attempted to persuade his fellow biologists to reconstitute their field, to alter their actions or their dispositions to act.

To call these intellectual activities rhetoric of science, then, is only to register a claim already staked and mined; to view these apparently distinct enterprises as rhetoric is merely to make available to all a coherent tradition, a set of well-used intellectual tools.

Rhetoric of science differs from literature and science, a branch of study that also "stars" its texts. The texts privileged by literature and science are traditionally literary; the science of an era is studied for its ability to illuminate the literary productions of that era: Katherine Hayles's *The Cosmic Web* trains the concepts of scientific field theory on a set of contemporary novels influenced by this theory. In contrast, rhetoric of science proposes by means of rhetorical analysis to increase our understanding of science, both in itself and as a component of an intellectual and social climate. From this perspective, when Gillian Beer studies the impact of Darwin on Victorian intellectual life, she is doing not literature and science but rhetoric of science.

To say that a rhetoric of science views its texts as rhetorical objects, designed to persuade, is not to deny that there is an aesthetic dimension to science. From a rhetorical point of view, however, this dimension can never be an end in itself; it is always a means of persuasion, a way of convincing scientists that some particular science is correct. In science, beauty is not enough: Descartes's physics is beautiful still, but it is not still physics.

Rhetoric Applied to Scientific Texts

In a neo-Aristotelian rhetoric of science, the apparatus of classical rhetoric must be generally applicable; a formulation must be developed that is recognizably classical and, at the same time, a theory of the constitution of scientific texts. This is not to say that classical ideas of style, arrangement, and invention must be mapped point for point onto these texts. The notion is not that science is oratory; but that, like oratory, science is a rhetorical enterprise, centered on persuasion. Instead of searching for exact correspondences, we must, as we proceed, achieve a general sense that the categories of classical rhetoric can explain the observable features of scientific texts.

This task is made easier by the existence of a long tradition of rhetoric and rhetorical analysis. Classical rhetoric was never a unitary system, nor was the rhetorical tradition unified throughout its history: Aristotle, Cicero, Quintilian differ, as do Campbell, Whately, and Blair. Doubtless, if more texts survived, even more disagreement among classical authors would be evident.

But it is the continuities in the rhetorical tradition that are most striking, continuities that subsist generally throughout medieval and modern rhetoric. Writers still find arguments where classical orators found them; the organization of writing still owes a debt to classical ideas of arrrangement; and rhetoricians still think of style in terms that are largely classical. When young people learn to write, they still learn what Quintilian taught.

The rhetorical analysis of science is made plausible by the close connection between science and rhetoric in the ancient world. Early Greek thought concerning the material world fluctuated wildly. To Thales the fundamental substance was water; to Anaximenes it was air. To Heraclitus all was flux; to Parmenides change was an illusion. To this *embarras de richesses* there were two reactions. The first was to ensure the certainty of knowledge; this was the way of Plato and Aristotle. The second was to regard knowledge as human and changeable, as rhetoric; this was the way of the Sophists.

The problem that rhetoric of science addresses, then, was set early in the intellectual history of the West. And then as now, this problem cannot be addressed unless rhetorical analysis includes not only the style and arrangement of science, but also those of its features usually regarded as unrhetorical—features commonly construed not as rhetoric but as the discovery of scientific facts and theories. From the rhetorical point of view, scientific discovery is properly described as invention.

Why redescribe discovery as invention? To discover is to find out what is already there. But discovery is not a description of what scientists do; it is a hidden metaphor that begs the question of the certainty of scientific knowledge. Discovered knowledge is certain because, like America, it was always there. To call scientific theories inventions, therefore, is to challenge the intellectual privilege and authority of science. Discovery is an honorific, not a descriptive term; and it is used in a manner at odds with the history of science—a history, for the most part, of mistaken theories—and at odds with its current practice, a record, by and large, of error and misdirection. The term *invention,* on the other hand, captures the historically contingent and radically uncertain character of all scientific claims, even the most successful. If scientific theories are discoveries, their unfailing obsolescence is difficult to explain; if these theories are rhetorical inventions, no explanation of their radical vulnerability is necessary.[2]

Stasis

At any time, in any science, scientists must make up their minds about what needs to be explained, what constitutes an explanation, and how such an explanation constrains what counts as evidence. When scientists think about matters of explanation, they are deciding what it is to do science. In rhetorical terms, they are using *stasis* theory, which is an established part of invention: a set of questions by means of which we can orient ourselves in situations that call for a persuasive response. In courtroom arguments, we consider whether an act was committed (*an sit*); whether it was a crime (*quid sit*); whether the crime is justified in some way (*quale sit*). In the analysis of law, these *stases* have a central role; in the analysis of science, their centrality is equally apparent.

1. *An Sit.* In the sciences, what entities really exist? Does phlogiston? Do quarks? Before Einstein's papers on Brownian movement, the existence of atoms was in question; afterward, their existence was regarded as confirmed.

2. *Quid Sit.* Given that certain entities exist, what is their exact

character? From antiquity, light has been steadily the subject of scientific scrutiny. Is light Aristotle's alteration in a medium, Descartes's pressure, Newton's particle, Young's wave (another alteration in a medium), or the zero-mass particle of quantum electrodynamics?

3. *Quale Sit.* Even if the character of an entity or phenomenon remains roughly the same, the laws governing it may be radically different: the same law of refraction that is, for Newton, the result of deterministic forces acting on minute particles is, for Feynman, the product of probabilistic ones acting on zero-mass particles.

Particular scientific texts emphasize particular *stases*. Although Einstein incidentally established the physical existence of atoms, he was mainly concerned with the *quale sit* of Brownian motion. Papers in evolutionary taxonomy establishing a new species mainly support the *quid sit* of existence, but they are also concerned with the *quale sit* of evolutionary theory. In every case the *stases* focus the scientist's attention on a particular aspect of the problem before him: Newton and Descartes, for instance, were both concerned with the nature of white light, the *quid sit*.

There is a final *stasis* applicable to both rhetoric and science: whether a particular court has jurisdiction. Whether something is a scientific theory depends on who is doing the judging. Newton's formulation of the theory of light remained the same throughout his career. At first it was rejected; its later acceptance did not depend on any alteration of the theory but on a change of jurisdiction. In the first court of opinion, Newton was judged by rules of others' making; in the final court, the rules were Newton's own.

Jurisdiction is also important in adjudicating the relationships between science and society. At what point do decisions cease to be internal to a science? The Inquisition saw itself as an appropriate arbiter of all knowledge, including Galileo's scientific theories. In modern times this determination is usually, and rightly, made by the scientific community. But even contemporary American courts see themselves as proper judges of the social impact of recombinant DNA research.

At any one time, in any one science, there are proper and improper ways to respond to the first three *stases*. For Aristotle, for example, the phenomena in need of explanation are those that naturally present themselves; what accounts for the motion of a stone released from the thrower's hand? This is a case of violent motion, a movement whose efficient cause is the application of a force to an object, overcoming its material cause, its *gravitas*. There is no violent motion without direct contact: a stone thrown in the air continues its motion after it leaves the thrower's hand only because of the impulsive power of the air directly behind it. The stone's initial trajectory is the formal cause of this violent motion. At the height of that trajectory, natural motion takes over; the stone begins to fall, seeking its natural place, the final cause of its motion. The material cause is again the stone's *gravitas*, its formal cause the downward trajectory itself, its efficient cause the distance from its natural place. For Aristotle scientific explanation is essentially qualitative, according to the four causes; mathematics has no place in physics.

In his *Principia*, Newton escapes the Aristotelian *an sit;* he no longer takes as *explanandum* the traditional topic of motion. For Newton it is not motion but change of motion that requires explanation. Motion itself—intuitively, the natural puzzle—is an *explanandum* no longer in the realm of science. Moreover, Newtonian explanations that do not enumerate all four causes can still be scientific. The material cause of change of motion is largely bracketed, and its final cause assigned to theology. The efficient and formal causes are privileged, and the formal cause is given a mathematical interpretation: change of motion is explained according to strict mathematical relationships among such nonobservables as force and mass. Such relationships permit quantitative solutions to problems in physics. Wherever possible, these problems have an experimental realization: in principle, though not in fact, Newton will assert only what he can observe under experimental constraints or can infer directly from controlled observation.

Because the presuppositions of Aristotle and Newton were opposed, because their notions of evidence and explanation seriously diverged, the sciences they created differed radically. Differently interpreted, the *stases* can lead—in fact, have led—to radically different conceptions of science. Since they precede science, the province of these interpretations cannot be science; their proper province is rhetoric.

Logos

The common topics are a staple of classical rhetorical invention; comparison, cause, definition—these and their fellows are the traditional places where rhetoricians can find arguments on any given topic. These same common topics are also an important source for arguments in science—in Newton, for example. In his *Opticks*, Newton defines a light ray twice. Early in this work he provides a definition in terms of the observable: light behaves *as if* it were made up of tiny particles. Later Newton defines light in terms of a hypothesis concerning the constitution of matter: light *may actually consist* of tiny particles. The difference in these definitions reflects a change in persuasive purpose. By means of the first definition, Newton hopes to persuade the skeptical scientist of the truth of his analysis of light; to agree, this scientist need not subscribe to Newton's speculative atomism. By means of the second, Newton hopes that this same scientist will seriously entertain atomism as a scientific hypothesis.

In Newton's optical works, the common topics are used heuristically as well as persuasively. Newton undermines Descartes's analysis of color by means of the topic of comparison: he contrasts Descartes's theory with his own incontrovertible experimental results. Concerned about the material constitution of light, he addresses the topic of cause: the sensation of light, he speculates, is evoked when its tiny particles impinge on the retina. In his presumption of the rectilinear propagation of light, he relies on the topic of authority; everybody since Aristotle has taken this as true.

In each case, we might say that Newton defines scientifically, compares scientifically. But in none of these instances is it possible to define a scientific sense for the common topics that is qualitatively distinct from their rhetorical sense: these sources for arguments in science and rhetoric do not differ in kind.

In addition to the common topics suitable to all argument, there are special topics that provide sources of argument for each of the three genres of speeches: forensic, deliberative, and epideictic. Forensic texts establish past fact; they are so named because their paradigm is the legal brief; their special topics are justice and injustice. Epideictic texts celebrate or calumniate events or persons of importance; their paradigms are the funeral oration and the philippic; their special topics are virtue and vice. Deliberative texts establish future policy; their paradigm is the political speech; their special topics are the advantageous and the disadvantageous, the worthy and the unworthy.

Scientific texts participate in each of these genres. A scientific report is forensic because it reconstructs past science in a way most likely to support its claims; it is deliberative because it intends to direct future research; it is epideictic because it is a celebration of appropriate methods. Analogously, scientific textbooks strive to incorporate all useful past science, to determine directions for future research, and to commend accepted methods. But science also has special topics of its own, unique sources for its arguments. Precise observation and prediction are the special topics of the experimental sciences; mathematicization is the special topic of the theoretical sciences. But there is considerable reciprocity. In the experimental sciences, mathematization is also a topic, and it provides arguments of the highest status; and in the theoretical sciences, at least by implication, arguments from mathematics are anchored in the special topics of prediction and observation.

But are observation, prediction, and mathematization *topics*? Science is an activity largely devoted to the fit between theories and their brute facts; the better the fit, the better the science. Surely, observation, prediction, and mathematization are not topics, but means to that end. In prediction, the confrontation between theory and brute fact is at its most dramatic. Einstein's theory of general relativity forecast the never-before observed bending of light in a gravitational field; Crick's theory of the genetic code predicted that an otherwise plausible variant—the codon UUU—would never occur. Both predictions insisted on the participation of nature; nature, not human beings, would clinch the argument. Einstein's theory was confirmed by the bending of stellar light as measured during a total eclipse; Crick's was disconfirmed by the discovery of a UUU codon. In both cases, it seems, we have left rhetoric behind. We seem to be in direct contact with the brute facts as the criterion for theoretical truth: stellar photographs in the first case, instrument readings in the second.

But this line of argument fails: in neither case did the brute facts point unequivocally in a particular theoretical direction. In fact, in no scientific case do uninterpreted brute facts—stellar positions, test-tube residues—confirm or disconfirm theories. The brute facts of science are stellar positions or test-tube residues *under a certain description;* and it is these descriptions that constitute meaning in the sciences. That there are brute facts unequivocally supportive of a particular theory, that at some point decisive contact is made between a theory and the naked reality whose working it accurately depicts, is a rhetorical, not a scientific, conviction. Observation, prediction, measurement, and their mathematization: these are sources for the arguments in science in the same way—exactly the same way—that the virtuous is the source of arguments for the epideictic orator.

The Structure of Argument. For Aristotle, scientific deduction differs in kind from its rhetorical counterpart. True, both are conducted according to the "laws" of thought. But rhetorical deduction is inferior for two reasons: it starts with uncertain premises, and it is enthymematic: it must rely on an audience to supply missing premises and conclusions. Since conclusions cannot be more certain than their premises, and since any argument is deficient in rigor that relies on audience participation for its completion, rhetorical deductions can yield, at best, only plausible conclusions. Rhetorical induction, reasoning from examples, is equally marked by Aristotle as inferior to its scientific counterpart because of its acknowledged inability to guarantee the certainty of its generalizations: examples illustrate rather than prove.

Aristotle notwithstanding, rhetorical and scientific reasoning differ not in kind but only in degree. No inductions can be justified with rigor: all commit the fallacy of affirming the consequent; as a result, all experimental generalizations illustrate reasoning by example. Deductive certainty is equally a chimera; it would require the uniform application of laws of thought, true in all possible worlds; the availability of certain premises; and the complete enumeration of deductive chains. But of no rule of logic—not even the "law" of contradiction—can we say that it applies in all possible worlds. Moreover, even were such universal rules available, they would operate not on certain premises but on stipulations and inductive generalizations. In addition, all deductive systems are enthymematic: the incompleteness of rhetorical deduction is different only in degree, not in kind, from the incompleteness of scientific deduction. No deductive logic is a closed system, all of whose premises can be stipulated; every deductive chain consists of a finite number of steps between each of which an infinite number may be intercalated (Davis and Hersh 1986, pp. 57–73). Because the logics of science and rhetoric differ only in degree, both are appropriate objects for rhetorical analysis.

Ethos and *Pathos*

Scientists are not persuaded by *logos* alone; science is no exception to the rule that the persuasive effect of authority, of *ethos*, weighs heavily.

The antiauthoritarian stance, the Galilean myth canonizing deviance, ought not to blind us to the pervasiveness of *ethos*, the burden of authority, as a source of scientific conviction.[3] Indeed, the progress of science may be viewed as a dialectical contest between the authority sedimented in the training of scientists, an authority reinforced by social sanctions, and the innovative initiatives without which no scientist will be rewarded.

Innovation is the *raison d'être* of the scientific paper; yet in no other place is the structure of scientific authority more clearly revealed. By invoking the authority of past results, the initial sections of scientific papers argue for the importance and relevance of the current investigation; by citing the authority of past procedure, these sections establish the scientist's credibility as an investigator. All scientific papers, moreover, are embedded in a network of authority relationships: publication in a respected journal; behind that publication, a series of grants given to scientists connected with a well-respected research institution; within the text, a trail of citations highlighting the paper as the latest result of a vital and ongoing research program. Without this authoritative scaffolding, the innovative core of these papers—their sections on results, and their discussions—would be devoid of significance.

At times, the effects of scientific authority can be stultifying: collective intellectual inertia blocked the reception of heliocentric astronomy for more than a century; Newton's posthumous authority retarded the reemergence of the wave theory of light. At other times, perhaps more frequently, authority and innovation interact beneficially; consider heliocentric astronomy between Copernicus and Kepler, the theory of light between Descartes and Newton, the concept of evolution in Darwin's early thought: in each of these cases we can see the positive results of the dialectical contest between authority and innovation. These examples alert us to the fact that there is no necessary conflict between originality and deference. One of the persuasive messages of authority in science is the need to exceed authority; indeed, the most precious inheritance of science is the means by which its authority may be fruitfully exceeded: "Was du ererbt von deinen Vätern hast / Erwirb es, um es zu besitzen" ("You must earn what you inherit from your fathers; you must make it your own"; Goethe, quoted in Freud 1949, p. 123).

At the root of authority within science is the relationship of master to disciple. To become a scientist is to work under men and women who are already scientists; to become a scientific authority is to submit for an extended period to existing authorities. These authorities embody in their work and thought whatever of past thought and practice is deemed worthwhile; at the same time, they are exemplars of current thought and practice. In their lectures, they say what should be said; in their laboratories, they do what should be done; in their papers, they write what should be written.

As long as science is taught as a craft, through extended apprenticeship, its routes to knowledge will be influenced by the relationships between masters and disciples. The modern history of heliocentricity is one of progress from epicycles to ellipses. But this theoretical development was realized only through a chain of masters and disciples, surrogate fathers and adopted sons: Copernicus and Rheticus, Maestlin and Kepler. By this means, research traditions are founded, and the methodological and epistemological norms that determine the legitimacy of arguments are passed on as tacit knowledge.

An examination of the forms of authority within science reminds us that epistemological and methodological issues cannot be separated from the social context in which they arise: the early members of the Royal Society decided what science was, how it would be accomplished, how validated, how rewarded. But we need also to be reminded of another set of authority relationships: those between science and society at large. It was the paradoxical promise of early science that it would benefit society best when wholly insulated from larger social concerns. This ideological tenet becomes difficult to justify, however, in an age of nuclear power and gene recombination. Justification is especially difficult when science converts its exceptional prestige into a political tool to protect its special interests, perhaps at the expense of the general interest. The recombinant DNA controversy is a case in point.

Emotional appeals are clearly present in the social interactions of which science is the product. In fact, an examination of these interactions reveals the prominent use of such appeals: the emotions are plainly involved, for instance, in peer review procedures and in priority disputes. Anger and indignation are harnessed in the interest of a particular claim; they are part of the machinery of persuasion. When science is under attack, in cases of proposed research in controversial areas, emotional appeals become central. The instance of proposed research in gene recombination is a good example of the fundamental involvement of science in issues of public policy, and of the deep commitment of scientists to a particular social ideology.

In addition, the general freedom of scientific prose from emotional appeal must be understood not as neutrality but as a deliberate abstinence: the assertion of a value. The objectivity of scientific prose is a carefully crafted rhetorical invention, a nonrational appeal to the authority of reason; scientific reports are the product of verbal choices designed to capitalize on the attractiveness of an enterprise that embodies a convenient myth, a myth in which, apparently, reason has subjugated the passions. But the disciplined denial of emotion in science is only a tribute to our passionate investment in its methods and goals.

In any case, the denial of emotional appeal is imperfectly reflected in the scientific texts themselves. The emotions, so prominent in peer review documents and in priority disputes, are no less insistently present in scientific papers, though far less prominent. In their first paper Watson and Crick say of their DNA model that it "has novel features which are of considerable biological interest" (1953b, p. 737). In his paper on the convertibility of mass and energy, Einstein says: "It is not impossible that with bodies whose energy-content is variable to a high degree (e.g. with radium salts) the theory may be successfully put to the test" (1952, pp. 67–71). In these sentences, key words and phrases—"novel," "interest," "successfully," "put to the test"— retain their ordinary connotations. Moreover, in Watson and Crick, "considerable" is clearly an understatement: the topic is the discovery of the structure of the molecule that controls the genetic fate of all living organisms.

Our science is a uniquely European product barely three centuries old, a product whose rise depended on a refocusing of our general interests and values. Its wellspring was the widening conviction that the eventualities of the natural order depended primarily not on supernatural or human intervention but on the operation of fixed laws whose preferred avenue of discovery and verification was quantified sensory experience. The ontological results of this epistemological preference defined the essence of nature and founded a central Western task: to control nature through an understanding of its laws. To this task, the specific values of science—such as the Mertonian norms of universalism and organized skepticism—are instrumentally subordinate. Equally subordinate are the values on which theory choice depends: simplicity, elegance, power. In such a view, *ethos*, *pathos*, and *logos* are naturally present in scientific texts: as a fully human enterprise, science can constrain, but hardly eliminate, the full range of persuasive choices on the part of its participants.

Arrangement

In science, the arrangement of arguments is given short shrift. It is hardly noticed, never taught; yet arrangement has always been important in modern science. Realizing its powerful effect, Newton cast his physics and recast his optics in Euclidian form. Indeed, during the three centuries of modern science, arrangement has become more, rather than less, important; more, rather than less, rigid. Currently, form is so vital a component that no paper can be published that does not adhere closely to formal rules. In fact, the arrangement of scientific papers has become so inflexible that even experienced scientists occasionally chafe under its restrictive principles: results in this section, discussion in that. But when P. D. Medawar, a scientist of wide influence, put his Nobel weight behind a mild reform—putting the discussion section first—his arguments were ignored rather than answered (1964, pp. 42–43).

Yet nothing is more artificial than the form of scientific papers. Experimental papers, for example, are not so much reports as enactments of the ideological norm of experimental science: the unproblematic progress from laboratory results to natural processes. It is of no consequence that such progress is far from unproblematic, or that the philosophical bases of this version of the scientific method have long been undermined. In experimental reports, arrangement is regarded as a sacred given.

There is another aspect of arrangement, one even more central to the operation of science. Aristotle's decision to privilege the proofs of logic and mathematics, to except them from the province of rhetoric, was itself rhetorical; it was a decision in favor of certain arrangements, a choice that rested on their presumed correspondence to the laws of thought. It is a truism that logical and mathematical proofs are purely matters of syntax, of form, austere tributes to the power of pattern to evoke the impression of inevitability:

All A is B
All C is A
All C is B

Like all syllogisms, this paradigm syllogism of science is sound only by virtue of its form, its arrangement. But so paradigmatic of absolute conviction have the forms of logic become, so binding has logical necessity seemed, that its force has been attributed to arguments in the natural sciences and, even, in the humanities: we speak of physical necessity and moral necessity, as if they and logical necessity were precisely analogous (Perelman and Obrechts-Tyteca 1971, pp. 193–260).

Style

From the beginning, stylistic choices in modern science have been deliberately trivialized: in the words of Bishop Sprat, the first historian of modern science, its communications must "return back to the primitive purity and shortness, when men delivered so many things in an equal number of words" (Sprat 1667, II, pp. xx). In such a program, the schemes and tropes of classical rhetoric are rigorously to be avoided. Nouns stand for natural kinds; predicates for natural processes. Syntax, the structure of the sentence, is only the reflection of reality, the structure of nature.

Scientific style remains oxymoronic at its core: modest in its verbal resources, heroic in its aim—nothing less than the description of reality. Accordingly, tropes like irony and hyperbole are barred; they draw attention away from the working of nature. Stylistic devices like metaphor and analogy likewise cannot be condoned; they undercut a semantics of identity between words and things. Should scientific prose favor the active or the passive voice? This quarrel over schemes—over the appropriate surface subject of scientific sentences—masks essential agreement among the antagonists. Regardless of surface features, at its deepest semantic and syntactic levels scientific prose requires an agent passive before the only real agent, nature itself. By means of its patterned and principled verbal choices, science begs the ontological question: through style its prose creates our sense that science is describing a reality independent of its linguistic formulations.

Despite these strictures, tropes like irony and hyperbole do appear regularly in scientific reports, belying the alleged reportorial nature of these texts and underscoring their true, persuasive purpose. Although the official view is that metaphor and analogy have only a heuristic function, that they wither to insignificance as theories progress, tropes are central to the scientific enterprise, and never disappear altogether. In the *Origin of Species,* for example, a central argument is the analogy between artificial breeding and natural selection. This analogy was not abandoned as the theory matured; instead, it was the means by which the theory has been maintained and extended. Analogy is also central to the whole enterprise of experimental science: laboratory experiments are scientifically credible only if there is a positive analogy between laboratory events and processes in nature.

In sum, in science arrangement has an epistemological task, style an ontological one.

Aristotelian Rhetoric Updated

To practice the rhetoric of science, then, is to make the *Rhetoric* the master guide to the exegesis of scientific texts. To perform this task effectively, the *Rhetoric* must be updated. The achievements of those squarely in the rhetorical tradition are the easiest candidates for incorporation into a neo-Aristotelian rhetoric of science. Of these, Chaim Perelman's work is most nearly central. His masterpiece, *The New Rhetoric,* written in collaboration with L. Olbrechts-Tyteca, has as its strategic aim the rehabilitation of rhetoric as a discipline whose task is the analysis of persuasion in the humanities and the human sciences. Although Perelman does not deal with the natural sciences, the analysis of these is a plausible extension of the scope of his theory.

One central New Rhetorical concept useful in the analysis of science is the "universal audience," an ideal aggregate that can refuse a rhetor's conclusions only on pain of irrationality. Although the universal audience has been attacked as an ontological category, there is no disagreement that its assumption is a valid rhetorical technique (Johnstone 1978, pp. 101–106). Indeed, it is a technique essential to the sciences. The real audiences for papers in taxonomy and theoretical physics are vastly different in their professional presuppositions; nevertheless, all scientists attribute to imagined colleagues standards of judgment presumed to be universal: not in the sense that everyone judges by means of them, but in the sense that anyone, having undergone scientific training, must presuppose them as a matter of course.

There is a more sweeping, and more telling, criticism of *The New Rhetoric,* the accusation that Perelman and Olbrechts-Tyteca are seriously derelict in their philosophical duty: "One is never sure whether the authors are thinking of rhetoric primarily as a technique or primarily as a mode of truth. One wonders, too, what status the authors are claiming for the book itself" (Johnstone 1978, p. 99). This criticism is a reminder to all of us to take an unequivocal stand on the epistemological status of our own inquiries. In my work, I view the techniques of rhetoric expounded by Perelman and Olbrechts-Tyteca, techniques such as analogy, as the means by which we are persuaded that any mode of inquiry, including that of science, is a mode of truth.

A neo-Aristotelian theory of rhetoric should also be prepared to incorporate the results of relevant modern thinkers, those who purport to reveal through their work enduring qualitative patterns that undergird apparently unique verbal behavior. In rhetoric, Aristotle finds three persuasive appeals, three levels of rhetorical analysis. In an analogous fashion, the Russian formalist Vladimir Propp finds that the dramatis personae of fairy tales exhibit thirty-one functions exercised in seven spheres of action; Freud divides the mind's functions into ego, superego, and id; Jürgen Habermas analyzes speech acts by means of their relationship to their validity claims, to their communicative functions, and to reality.

The incorporation of views as divergent as those of Propp, Freud, and Habermas into a neo-Aristotelian rhetoric of science necessitates the abandonment of strong ontological claims. Aristotle's psychology and that of Freud cannot be incorporated into a single coherent theory. In addition, an explanatory pattern in which we put great store may be, from another, equally legitimate, point of view, epiphenomenal, a symptom of the operation of purportedly more fundamental processes: Propp's patterns may be an effect of Freudian imperatives; Freudian imperatives, a result of the social dynamics of the upper-middle-class Viennese Jews who were Freud's contemporaries. Our choice among these patterns must be based not on their relative truth, a judgment we cannot make, but on the amount each contributes to the understanding of the ways in which rhetorical processes constitute science.

In his *Crisis of European Sciences* Edmund Husserl highlights the success of the natural sciences, a success to be contrasted with the general failure of reason in its task of improving the everyday world, the moral, mental, social, and physical space that all human beings share. Husserl locates this failure in the rupture caused by the dualism of Descartes. Whatever its source, the breach between the world of science and our human world is real enough, and the task of reconciliation is as pressing today as it was for Husserl. Because it sees science wholly as a product of human interaction, rhetoric of science is a gesture in the direction of such reconciliation, an argument for the permanent bond that must exist between science and human needs.

The question of whether rhetorical analysis is appropriate and equal to so formidable a task arises not as a consequence of any eternal truth or reasoned argument, but only as a result of the progressive narrowing and devaluation of rhetorical studies since Plato. It was Plato's successful attack on the Sophists that separated rhetoric from truth; it was the long authoritarian winter of the Roman Empire that limited rhetoric to its forensic and epideictic forms; it was the sterile intellectual reformulation of Ramus that reduced rhetoric to matters of style. That this narrowing was equally a degradation can be seen in phrases such as "mere rhetoric" or "empty rhetoric."

Turning our backs on this past, we can engage in a systematic examination of the most socially privileged communications in our society: the texts that are the chief vehicles through which scientific knowledge is created and disseminated. We can argue that scientific knowledge is not special, but social; the result not of revelation, but of persuasion. In this way we can see science as a permanent component of Husserl's life-world, where it has its origin, and from which it must obtain all its purpose.

CHAPTER 2

Analogy in Science

Chaim Perelman asserts that science "elaborates a system of necessary propositions which will impose itself on every human being, concerning which agreement is inevitable" (1971, p. 2). This opinion falls easily within the tradition of Aristotle, for whom the scientist is "proof against the persuasion of argument" (1960b, p. 519). To those with this view, scientific knowledge is privileged because it rests on a firm foundation—the bedrock of reality beneath a world of appearance. So unswerving a conviction could survive neither the no-man's-land of twentieth-century physics nor the rigors of twentieth-century philosophy. Indeed, this absolutist view of scientific truth now has an alternative, a sophisticated relativism in which truth depends not on conformity to a substratum of reality, but on agreement among significant persons. No theory of rhetoric in science can ignore this generally plausible relativism: since all truth is intersubjective, science, like all persuasive discourse, must convince us of the truth of its claims.

From a rhetorical point of view, the high esteem bestowed upon science gives its communications a built-in *ethos* of especial intensity. Ziman says that science has "tremendous *rhetorical* power . . . overwhelming persuasive force" (1968, p. 31); Douglas calls science "the most powerful rhetoric of all today" (1971, p. 57; see also Weigert 1970, Overington 1977, Brummett 1976, Kelso 1980). In this chapter I will clarify this special power of scientific rhetoric by focusing on some differences in the communication strategies of political oratory, scholarly argument, and scientific reports, differences based on the way these disciplinary domains use analogy.

I have chosen analogy for scrutiny because it is pervasive and revelatory. Analogy is traditionally a device central to rhetorical proof; its continuing importance in modern oratory is illustrated in Churchill's dictum that "apt analogies are among the most formidable weapons of the rhetorician" (quoted in Montalbo 1978, p. 6). As Perelman's long section on analogy reminds us, this method of argument is also pervasive in philosophy (pp. 371–459). Finally, analogy has had a long history in the sciences. An important device in Aristotle's scientific writing, it is still very much in use today: the concept of the genetic code is a scientific analogy.[1] I try to show that each domain—oratory, science, and scholarship—uses analogy differently and that these differences reflect distinctions in the kinds of intersubjective agreement sought and achieved. I also hope to show that each domain has a character and a "truth" of its own, that the impressive closure achieved by the sciences is as much a rhetorical as an epistemological triumph.

Analogy in Political Oratory

An excellent example of the use of analogy in political oratory is the work of an acknowledged master, Franklin Delano Roosevelt. On March 4, 1933, FDR delivered his First Inaugural to a nation in the grip of the Great Depression. A single analogy governs the speech:

the tenor	the vehicle
a President	*a general*
the Depression	*a foreign invasion*

Roosevelt introduces this military analogy when he speaks of "terror which paralyzes needed efforts to convert retreat into advance." Later, he refers to the task ahead as "the emergency of a war." The American people, he says, "must move as a trained and loyal army willing to sacrifice for the good of a common discipline," and he asks for "a unity of duty hitherto evoked only in time of armed strife":

I am prepared under my constitutional duty to recommend the measures that a stricken nation in the midst of a stricken world may require.
These measures, or such other measures as the Congress may build

out of its experience and wisdom, I shall seek, within my constitutional authority, to bring to speedy adoption.

But in the event that the Congress shall fail to take one of these two courses, and in the event that the national emergency is still critical, I shall not evade the clear course of duty that will then confront me.

I shall ask the Congress for the one remaining instrument to meet the crisis—broad executive power to wage war against the emergency, as great as the power that would be given to me if we were in fact invaded by a foreign foe. (1963, p. 198)

In the First Inaugural the appeal to reason is firmly minimized, as Eleanor Roosevelt discovered to her dismay. She found the Inaugural terrifying "because when Franklin got to that part of his speech when he said it might become necessary for him to assume powers ordinarily granted to a President in war time, he received his biggest demonstration." That Roosevelt's audience clearly preferred his strong leadership is indicated by the nearly half a million letters the White House received immediately following the speech (Schlesinger 1959, p. 1).

The minimization of rational appeal was confirmed at Roosevelt's first news conference four days later. FDR showed that, although his bid for strong leadership was clear, his programs were vague indeed:

QUESTION: In your inaugural address in which you only touched upon things, you said that you are for sound and adequate [currency]—
THE PRESIDENT: I put it the other way around. I said "adequate but sound."
QUESTION: Now that you have more time, can you define what that is?
THE PRESIDENT: No. (Laughter) In other words—and I should call this "off the record" information—you cannot define the thing too closely one way or the other. (1972, vol. 1, p. 11)

As one congressman complained, "He is for sound currency, but lots of it" (quoted in Leuchtenburg 1972, p.42).

This attitude toward intellectual analysis runs throughout the speech. Roosevelt's goal is not rational assent, based on "a genuine understanding of the position assented to" (Johnstone 1963, p. 92), but emotional commitment. The listener is asked to reconceive social reality, to see the civilian tenor entirely in terms of the military vehicle, unequivocally to prefer FDR's authoritarian stance.

Why is this analogy persuasive? In the general case, "the aptness of the equivalence invoked is the condition of its effectiveness." But aptness depends on understanding, and "understanding of the [vehicle of an analogy] . . . presupposes adequate knowledge of the place it occupies in a given culture" (Oakeshott 1962, p. 235; Perelman 1971, p. 391). In Roosevelt's analogy, understanding the effectiveness of the military vehicle means perceiving the natural fear of invasion on the part of a population only the oldest of whom could remember battles fought within its borders. Effectiveness depends also on fear of another sort of invasion, the terror of infection invoked by the adjective "stricken." To form a bridge between tenor and vehicle, between the Depression and foreign invasion, the analogy assumes the knowledge and acceptance of Roosevelt's dual constitutional role as President and Commander-in-Chief.

It would be wrong to call the American response irrational. Because we can seldom carefully consider our emotional responses, we tend to regard them as irrational and disruptive. Nevertheless, we may be truer to our experience if we see these responses as judgments of a special sort, judgments whose general purpose is to maximize self-esteem. Our general perception of the disruptiveness of strong emotional responses remains accurate; but this perception now arises entirely from the extraordinary exigencies that are the proximate cause of strong emotions, exigencies that exceed the carrying capacity of our usual repertoire of responses. This view makes the emotions rational and purposive, an aspect of our ethical behavior, an arena for learning (Solomon 1980, p. 270; pp. 251–52).

This analysis of the emotions helps to explain the American response to Roosevelt's First Inaugural. The response required was assent, whether or not it was based on genuine understanding; but the response was not in other respects less than rational. It was an emotional reaction to crisis, a reaction that enhanced self-esteem but did not exclude the acceptance of full responsibility for its consequences. Analogy was the instrument of this persuasion, an analogy that mobilized strong emotions in the interest of the wholesale transformation of social reality, a precursor to a new set of beliefs whose survival value was viewed as significantly higher than the old. As a result of this transformation, the military vehicle fused with the civilian tenor, creating a new theme, a *topos* for popular support of the strong executive measures that characterized the federal activity of Roosevelt's first hundred days.

Analogy in Scholarly Argument

For a good portion of their professional lives, Sir Karl Popper and Thomas S. Kuhn have been arguing with each other over the meaning of continuity and progress in the sciences. Kuhn believes that scientific revolutions—such as the Copernican—result in major changes in the way scientists perceive the world and formulate problems; thus scientific growth is discontinuous, not progressive, although "at least part of the [preceding] achievement always proves to be permanent" (1962, p. 25). Popper, on the other hand, believes that "in science (and only in science) can we say that we have made genuine progress: that we know more than we did before" (1970, p. 57).

In a passage from one of his essays, Popper uses analogy to assert the possibility of significant communication between rival major scientific systems ("frameworks"), such as the Ptolemaic and the Copernican. Without this possibility, he believes, there can be no scientific progress: "It is just a dogma—a dangerous dogma—that the different frameworks are like mutually untranslatable languages. The fact is that even totally different languages (like English and Hopi, or Chinese) are not untranslatable, and that there are many Hopis or Chinese who have learned to master English very well" (1970, p. 56).

In his response to Popper, Kuhn accepts his analogy but draws a crucial distinction between the ability to learn another language and the ability to translate from one language to another. This method of refutation is effective insofar as "it emphasizes the fragility and arbitrary character of the analogy" (Perelman 1971, p. 387).

> I accept the utility, indeed the importance, of the linguistic parallel, and shall therefore dwell for a bit upon it. Presumably Sir Karl accepts it too since he uses it. If he does, the dogma to which he objects is not that frameworks are like languages but that languages are untranslatable. But no one ever believed they were! What people have believed, and what makes the parallel important, is that the difficulties of learning a second language are different from and far less problematic than the difficulties of translation. Though one must know two languages in order to translate at all, and although translation can then always be managed up to a point, it can present grave difficulties to even the most adept bilingual. He must find the best available compromises between incompatible objectives. Nuances must be preserved but not at the price of sentences so long that communication breaks down. Literalness is

desirable but not if it demands introducing too many foreign words which must be separately discussed in a glossary or appendix. People deeply committed both to accuracy and to felicity of expression find translation painful, and some cannot do it at all. (1970, p. 267)

Like languages, rival scientific theories are untranslatable because they "cut up the world in different ways." Just as different languages divide the color spectrum differently—the Hindus, for instance, have one word for both orange and red—so in a science like chemistry key terms change their meanings after scientific revolutions: "alloys were compounds before Dalton, mixtures after" (1970, p. 269).

In this exchange between Popper and Kuhn, analogy is both the agent of intellectual exploration and a vehicle for proof. As a heuristic, it is active, helping to create hypotheses. Nevertheless, analogy also serves a probative purpose: the translation analogy of Popper and Kuhn is as much a part of their argument as Roosevelt's military analogy is part of his. In Roosevelt's case, however, the chief criterion for success is assent regardless of understanding, assent procured mainly through the emotional force of the military comparison. Popper and Kuhn's analogy, on the other hand, is successful only insofar as its force is rational: the assent it legitimates fully depends on the understanding it creates. This difference in aim is fostered by the emotional neutrality of the translation analogy, so different in nature from the emotionally charged comparison used by Roosevelt.

Not that scholarly discourse is free from emotional heat. Sir Karl's prose contains the admonitory locution "a dogma—a dangerous dogma." And Kuhn expresses exasperation: "But no one ever believed they were!" Neither of these locutions, however, is central to the argument. But if the emotions are peripheral to their argument, the personalities of the individual scholars are not. In fact, the premium placed on individual intellectual brilliance and the relative absence of unanimity and progress in the humanities and in the social sciences are not unconnected. Physics excludes neither individual brilliance nor acrimonious dissent, but physicists may infer what humanists may not: that the problems on which they are concentrating are connected in such a way that their joint solution will further knowledge.

In the philosophy of science, this feeling is an exception. Popper and Kuhn raised the level of discourse in philosophy of science to new heights. Nevertheless, the unity of a discipline and its progress depend not on the depth of individual intellectual achievement, but on significant agreement among its practitioners; its Poppers and its Kuhns must agree concerning the criteria by which significant theoretical differences can be resolved. And one senses that, in all likelihood, Popper and Kuhn will continue to disagree, that no evidence on the one side will be judged conclusive by the other. As a result, intellectual performance in philosophy of science will be spurred on by the need to refute the opponents' arguments or to support one's own, to gain victories, recruit disciples, or effect conversions.

Kuhn's rebuttal of Popper within the context of the latter's own analogy illuminates the limits of possible agreement in scholarly argument. Such arguments are not open to everyone; they can occur only among a small group with specialized prerequisites, men and women trained in the complex etiquette of a discipline, comfortable with its styles of reasoning and able to solve its current problems in a manner judged worthy by its intellectual guardians. In this sense Popper and Kuhn may be said to agree, for if assent to or dissent from their positions were not based on "genuine understanding" of these positions, their combats would be devoid of intellectual significance (Johnstone 1963, p. 92).

Nevertheless, this commitment to scrutinize analogical arguments defines the limits of significant agreement in fields that are scholarly rather than scientific. In science, though not in scholarship, this commitment is supplemented by a set of agreed-upon quantitative methods, methods that in the aggregate act as criteria for settling theoretical disputes. As a result, in science, in contrast to the humanities and social sciences, a reliable level of significant agreement is possible.

Analogy in Scientific Reports

Despite his best efforts, the seventeenth-century Swiss mathematician Jacques Bernoulli was unable to find the sum of the infinite series of the squares:

$$1 + 1/4 + 1/9 + 1/16 + 1/25 + 1/36 + 1/49 + \ldots = ?$$

Since the numbers in this series become progressively smaller, an approximate solution is readily available. In fact, in the following century another mathematician, Leonhard Euler, found an answer to seven places. Dissatisfied with this approximation, Euler, also a Swiss, made a daring leap by means of analogy: he used algebra, designed for finite mathematics, to solve an nonalgebraic equation, "applying the rules made for the finite to the infinite." This solution by analogy, and its accompanying answer of $\pi^2/6$, was in no way a proof. Finally, after much effort, Euler "succeeded in verifying . . . exactly, the value of $\pi^2/6$ for Jacques Bernoulli's series [by a proof that was] hidden and ingenious" (Polya 1954, p. 21). As this example demonstrates, analogy has a role in mathematics: not probative, but heuristic.

The "genetic code" is another example of the heuristic role of analogy. In the second quarter of the twentieth century, scientists repeatedly suggested a strict relationship between the inheritable characteristics of plants and animals and the exact structure and sequencing of genes within their chromosomes. The most famous statement of this relationship was an analogy. Erwin Schrödinger, the Nobel laureate in physics, in his speculative lecture series published as *What is Life?* makes the following assertion: "With the molecular picture of the gene it is no longer inconceivable that [a] miniature code should precisely correspond with a highly complicated and specified plan of development and should somehow contain the means to put it in operation" (1967, p. 66). In other words, there is an analogy: just as a code conveys information from one human being to another, a genetic code transfers genetic information from the hereditary substance to the protein that forms living matter.

Because of the virtual absence of observational and experimental evidence at the time, neither the alleged code nor its transmission mechanism could be elucidated, and statements like that of Schrödinger must be classed as speculation, not science. Not until 1953 did Schrödinger's genetic code analogy definitely prove its usefulness. In that year James Watson and Francis Crick made their Nobel Prize–winning discovery of the structure of the hereditary substance, now identified as deoxyribonucleic acid (DNA): "two helical chains each coiled round the same axis," each chain joined to its partner by a series of base pairs: "if an adenine [A] forms one member of the pair, on either chain, then on these assumptions, the other member must be thymine [T]; similarly for guanine [G] and cytosine [C]. The sequence of bases on a single chain does not appear to be restricted in any way. However, if only specific pairs of bases can be formed, it follows that if the sequence of bases on one chain is given, then the sequence on the other chain is automatically determined" (1953b, p. 738). Watson and Crick concluded "that the specific pairing we have postulated immediately suggests a possible copying mechanism for the genetic material" (1953b, p. 738). The nature of this "copying mechanism" remained to be elucidated.

It soon transpired that ribonucleic acid (RNA), closely related to DNA, was central to the elucidation of the genetic code. In RNA uracil [U], not thymine [T], is the fourth base, the fourth "letter." Within the decoding process, DNA transcribes the hereditary message onto RNA. By a complex series of mediations, RNA translates the message onto an appropriate protein, a message transcribed by units called codons.

Each codon or "word" is three bases long; each has to be "read" correctly in order for the hereditary mechanism to function properly. A correct reading is ensured if the genetic code is "comma-free." A comma-free code is so constructed that a correct reading is independent of starting place. Here is a lexicon of twenty codons for a comma-free triplet code (the slashes are for expository purposes only):

ACA/ACC/ACG/ACU/AGA/AGG/AGU/UCA/UCC/UCG/UCU/
AUA/AUU/UGA/UGG/UGU/GCA/GCC/GCG/GCU.
(Woese 1967, pp. 23–29)

To demonstrate the comma-free nature of this code, let us take the following imaginary short chain:

A	U	U	U	G	A	A	U	U	C	A
1	2	3	4	5	6	7	8	9	10	11	12

Given the above lexicon, there are only four codons in this chain: only the first, fourth, seventh, and tenth positions represent legitimate starts. A start at the second position, for instance, gives us the nonsense reading UUU.

Crick and his associates claimed that the genetic code was comma-free. But, despite its theoretical elegance and its ingenious interpretation of the coding analogy, their comma-free code did not stand up to mounting counterevidence. The final blow was rendered by Marshall W. Nirenberg—later a Nobel laureate—and his co-worker,

J. Heinrich Matthaei. They clearly established that the "addition of
. . . polyuridylic acid [to a stable cell-free system] specifically resulted
in the incorporation of L-phenylalanine into a protein resembling
poly-L-phenylalanine" (1961. p. 1601). Thus "at least one codon all
of whose bases were identical was assigned to an amino acid"(Woese
1967, p. 27)—the "illegitimate" codon (according to comma-free
rules) UUU. This meant that the correct reading of the genetic
substance was accomplished not by a comma-free code, but by accurate start and stop signals.

This historical example underlines the heuristic, as distinct from
the probative, use of analogy in the sciences.[2] The analogy that led
Crick to the correct interpretation of the structure of DNA also led
him to the incorrect formulation of the way the genetic code was
read. However, Nirenberg and Matthaei, working with the same
analogy, discovered the correct formulation.

Scientific reports and scholarly arguments are alike in the value
they place on the heuristic function of analogy and on the rules of
inference and evidence with which analogies and the hypotheses they
generate must be examined. However, scientific reports have
recourse to one additional tool: a complex of quantitative methodol-
ogies shared by scientists and central to their verification procedures.
For instance, Nirenberg and Matthaei's techniques of separation
(ultra centrifugation and electrophoresis) and of tracing (paper chro-
matography and radioactive isotopes) are well-founded technical procedures whose results can be challenged only by questioning the
conclusions of decades of work by thousands of scientists in dozens
of fields. As Black says: "In appraising [scientific] models as good or
bad, we need not rely on the sheerly pragmatic test of fruitfulness in
discovery; we can, in principle at least, determine the 'goodness' of
their 'fit'" (1962, p. 238).

Of course, the leap from quantitative readings to valid scientific
inferences is not automatic. In fact, because of the problematic nature
of this link, Nirenberg and Matthaei proceed with caution.[3] They
show that their unknown substance shares four key characteristics
with "authentic poly-L-phenylalanine" and conclude merely "that
polyuridylic acid contains information for the synthesis of a protein
having many of the characteristics of poly-L-phenylalanine." Still, the
example of the discovery of the genetic code demonstrates the rhetorical force of such methods of verification.

Crick heard about Nirenberg and Matthaei's devastating critique of his comma-free system at a Biochemical Congress in Moscow. He wrote: "The audience of Symposium [One] was startled by the announcement of Nirenberg that he and Matthaei had produced polyphenylalanine (that is, a polypeptide all the residues of which are phenylalanine) by adding polyuridylic acid (that is, an RNA the bases of which are all uracil) to a cell-free system which can synthesize protein. This implies that a sequence of uracils codes for phenylalanine" (Crick et al. 1961, p. 1232). This discovery forced Crick to reject his comma-free formulation in favor of the alternate hypothesis that "the correct choice [is] made by starting at a fixed point along the sequence of bases three (or four, or whatever) at a time." Such ready agreement demonstrates not so much fellow feeling as necessary assent to verification based on agreed-upon procedures.

That this effect is rhetorical is demonstrated by Chaim Perelman in his discussion of the dissociation of concepts. He speaks of "the 'appearance-reality' pair":

appearance	or, in general,	*term I*
reality		term II.

To Perelman, "term I corresponds to the apparent," while "term II provides a criterion . . . a rule that makes it possible to classify the multiple aspects of term I in a hierarchy. It enables those that do not correspond to the rule which *reality* [his emphasis] provides to be termed . . . erroneous." If this dissociation is applied to the case of the genetic code, the analogy becomes term I, the methods of verification, term II. (These are perhaps better known as the context of discovery and the context of verification.) Indeed, Perelman's insight into the characteristics of term II applies directly to my account of the discovery of the genetic code: "Term II profits from its oneness, from its coherence, when set against the multiplicity and incompatibility of the aspects of term I, some of which will be disqualified and marked ultimately for disappearance" (1971, pp. 416–417). The persuasive effect of science becomes just its ability to move from term I to term II *as if* moving from appearance to reality.

Political oratory, scholarly argument, scientific reports: in each, analogy functions differently. In political oratory, the path to assent is primarily through emotional commitment. In scholarly argument and in scientific reports, rational commitment is the primary route; but commitment in science reaches beyond argument and rests ultimately on agreed-upon procedures. Perhaps there is no scientific method, no global strategy for all of science; still, there are scientific methods, the aggregate of agreed-upon procedures. It is consensus over the necessity for these that, finally, differentiates reports in the sciences from political and scholarly discourse. Although each scientific procedure is doubtless based on argument, we can discover this only by means of a diligent search into the scientific past. For scientists, however, science has no past—or, rather, no past that does not wholly suit its present purposes. It is this absence, then, that nurtures the useful illusion: for scientists, the results of science depend not on argument but on nature herself.

CHAPTER 3

Taxonomic Language

A complete rhetoric of science must avoid this accusation: after analysis, something unrhetorical remains, a hard "scientific" core. In this chapter I want to test the hypothesis of completeness against evolutionary taxonomy, the science of classifying animals and plants as species in accordance with evolutionary theory. If a rhetoric of this science is possible, we must be able to reconstruct the central concept of evolutionary taxonomy, the species, rhetorically, without remainder. But our route to this goal must be indirect. We cannot translate the concept of species directly into rhetorical terms because we must avoid the claim that what is being translated is precisely what is *not* scientific, that after our rhetorical analysis there still remains an essential core of untranslatable scientific meaning.

To avoid this claim, we must reconstruct the species of evolutionary taxonomy rationally in the hope that the result will be recognizably a version both of this particular scientific concept and of "the methodological ideal . . . that dominates modern mathematically based natural science" (Gadamer 1975, p. 414); we must prepare a "schematized description of an imaginary procedure, consisting of rationally prescribed steps, which would lead to essentially the same results as the actual [historical or] psychological process" (Carnap 1963, p. 16). In a rational reconstruction of the species of evolutionary taxonomy, when stages are referred to, stages at which species are identified, defined, or redefined, these are not historical or psychological events but analytical categories. It is this rational reconstruction of the species that I will attempt to translate without remainder into rhetorical terms.

I do not deny that infinitely many rational reconstructions are possible; that evolutionary taxonomists may argue with the details of my rational reconstruction; that they may even take the position that mine is the wrong rational reconstruction. But I would argue that these scientists cannot take the position that rational reconstruction *per se* is the wrong approach for extracting method from practice without denying that evolutionary taxonomy is a science. And that is precisely my point. To evolutionary taxonomists, ours must be a world inhabited by evolutionary species. To each of these can be applied, from each of these can be inferred, at least in principle, a central process: the operation of natural selection on random genetic variation. Moreover, this reciprocal relationship between observation and theory presupposes objectivity: evolutionary taxonomy must describe a world whose existence is in some sense independent of its descriptions.

Our rhetorical reconstruction, on the other hand, redescribes the evolutionary species by translating into rhetoric each aspect of our rational reconstruction: no aspect we have described as science remains without its rhetorical counterpart. By means of rhetorical reconstruction, evolutionary taxonomy is transformed into an interlocking set of persuasive structures. *Sub specie rhetoricae,* we do not discover, we create: plants and animals are brought to life, raised to membership in a taxonomic group, and made to illustrate and generate evolutionary theory. If a rhetorical reconstruction describes rhetorically every aspect that a rational reconstruction describes rationally, a complete rhetoric of science becomes possible.

The Rational Reconstruction of Species: Identification

In contemporary biology, species cannot be defined classically, by genus and differentia; the effort, in Wittgenstein's pungent phraseology, resembles "[trying] to find the real artichoke by stripping it of its leaves." Instead, a species must be defined by depicting and describing "family resemblances" among organisms (1965, p. 125).[1] Earlier, in *Origin of Species,* Darwin had placed this notion in a taxonomical context: "There are crustaceans at the opposite ends of the series, which have hardly a character in common; yet the species at both ends, from being plainly allied to others, and these to others, and so onwards, can be recognized as unequivocally belonging to this, and to no other class of the Articulata" (1859, p. 419).

Family Resemblances

When we define general terms like species according to family resemblances, we give up as spurious the precision of classical definition as it applies to the natural world. Nevertheless, definition by family resemblance is real in the sense that it marks off a nonarbitrary class. Because I lack rules that will demarcate absolutely what I name as a new species of hummingbird, I can be puzzled about whether a particular organism belongs to the species, but I cannot just drop anything in there, eagles for example. In addition, a good general term is an open class in two senses: there is always the possibility that I will see a member I overlooked, and at least for creatures not extinct, there is always the chance of genuinely new members. Finally, the application of a general term can be learned, that is, a learner can apply the term independent of the guidance of his teacher (Bambrough 1966).

Potential Species

In a typical paper in evolutionary taxonomy, "A New Species of Hummingbird from Peru" (Fitzpatrick, Willard, and Terborgh 1979), there is everywhere evident the extensive description and depiction that are the evidential grounding of family resemblance. These scientists, for example, record numerous qualitative visual impressions: "A broad, pale buffy breast band separates the smaller throat spots from larger and more numerous discs on the breast and flanks. In a few specimens the posterior border of the breast band is entirely defined by a broad row of these discs. The belly is free of dark spots in all specimens. The downy crissum [anal region] is white as in males, and the undertail coverts [feathers at base] are dusky, edged Cinnamon" (1979, p. 178).

In addition to these visual impressions, Fitzpatrick and his co-workers define the potential species by recording its behavior. For example, we are afforded a detailed description of "male-female display": "Initially a pair [of birds] was foraging around the walls of [a vine-covered sinkhole] and in the surrounding shrubs along the rim. Both male and female were observed perching on a rootlet, making frequent sallies to capture tiny flying insects" (1979, p. 183).

To this abundance of qualitative information, quantitative data are added. We are given by degree, minute, and second the geographical coordinates of the area explored; to the nearest 10 meters the height of the habitat above sea level; to the tenth of a millimeter the mean, range, and standard deviation of six specimen dimensions; to two decimal places the ratio of exposed culmen (upper ridge of a bird's bill) to wing chord length; to the second the length of calls.

But the most impressive attempt to establish the species is in the frontispiece, a full-color painting of two of the birds in situ: a dimorphic pair, the female perched at an angle that most clearly displays her distinctive tail and underparts. The male displays his characteristic deeply cleft tail and dark coloring, and also the extension and rapid movement of his wings as he hovers. Moreover, the bird's feeding behavior is shown: the "deep purple petals [of its favorite flower] form a tubular corolla that hangs vertically . . . forcing the foraging [bird] to hover directly below and point its bill straight upward to retrieve the nectar" (Fitzpatrick, Willard, and Terborgh 1979, p. 182).

The potential species is not only described and depicted in terms of family resemblances; it is also, by comparison and contrast, carefully differentiated from closely allied species. In this process of establishing species status, the character is the minimum unit of observation. Characters are parts of the organism or aspects of its behavior that most perspicuously establish it as a member of a genus or differentiate it as a separate species (Mayr 1982, pp. 19–32). The harder a character works at either of these tasks, the more it is valued, the more heavily it is weighted.[2] In establishing their potential species as a member of a particular genus, Fitzpatrick and his co-workers emphasize "the relative bill length, nostril feathering, and well delineated nasal operculum [lid-like covering]"; in establishing the difference between the potential species and its fellows of the same genus, they give weight to the female's "elongated and deeply forked, entirely iridescent, metallic blue tail, equally bright on both surfaces; combined with buffy underparts interrupted by a pale pectoral [breast] band" (1979, pp. 177, 181).

Not just in the frontispiece but throughout this paper, pictures join words in the clear display of likenesses and differences, strongly reinforcing the presence of a potential new species. In Figure 1, for example, the seven drawings, considered in concert and without words, make equally emphatic the family resemblances crucial for

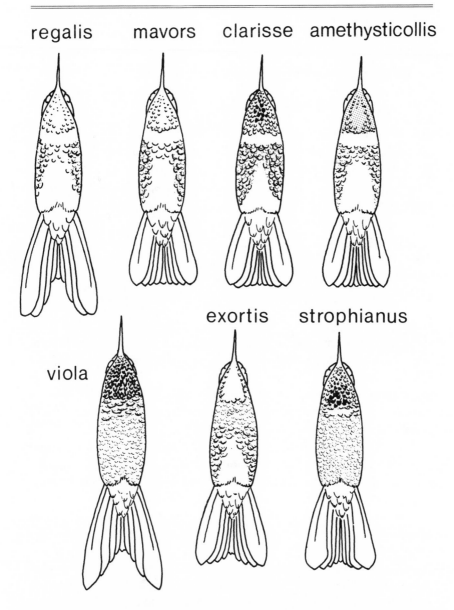

regalis   mavors   clarisse   amethysticollis

viola   exortis   strophianus

Figure 1 Ventral patterns and tail forms of *Heliangelus* females, including *H. regalis*. *H. micrastur* female resembles *exortis*, and *H. spencei* resembles *amethysticollis*. Note similarity between *regalis* and *mavors*. (From Fitzpatrick, Willard, and Terborgh 1979, p. 180. Reproduced by permission of *The Wilson Bulletin*.)

genus identity and the contrastive features essential for within-genus differentiation.

Taxonomic and Evolutionary Species

It is by means of statistical inference that this extraordinary detail is given taxonomic significance: in this way, Fitzpatrick and his co-workers definitely assert that they have a new species, a new living entity related to all such entities. Sixteen specimens have been collected:

	adult	sub-adult
male	5	5
female	5	1

Assuming a fair sample, the authors can describe the species by describing these specimens. Thus, when averages, ranges, and standard deviations are given, for example, those for wing chord length, they are equally those of the specimens collected and those of the species. Description, comparison, and statistical inference having done their work, the species can be officially certified as new by naming: it is a hummingbird, the genus *Heliangelus*, the species *regalis*, a Royal Sunangel.

But taxonomic identification is not the final stage: taxonomic species must be redefined in evolutionary terms. A species first characterized without Darwinian assumptions is now characterized in terms of a theory concerning the origin of species: that all stem from the operation of natural selection on random variations, some of which are favored because they increase the rate of survival of those who possess them.

In taxonomic work, evolutionary theory serves at least two purposes: it explains problematic observations that arise in the process of identification, and it licenses prediction. In the process of identifying their new species of hummingbird, Fitzpatrick and his collaborators make the following problematic observation. Among the species of *Heliangelus* to which *regalis* seems most closely related, the monochromatic plumage of the male *regalis* is strikingly anomalous. To account for this irregularity, the investigators invoke the concept of convergence (similarity in species resulting not from descent, but from environmental constraint):

> Thus, if *regalis* is indeed closest to this species as suggested by the appearance of the female, it has undergone a dramatic differentiation in which the male converged upon several more distant relatives. The elongated, narrow, metallic blue tail in both sexes of *regalis*, equally iridescent on both surfaces, is suggested only in *H. strophianus* (sexes similar) and *viola* (very large, dark green female lacks a breast band). (1979, p. 181)

But evolutionary theory does more than explain; it predicts. Prediction in evolutionary biology, of course, is not the same as in classical physics, where future events can often be forecast with startling quantitative accuracy. In evolutionary biology, although predictions need to be well-defined and consistent with the evidence on which they are based, they can legitimately vary in precision of measurement:

> For example, population ecologists are generally satisfied to explain to one order of magnitude the increases and decreases in population size of the organisms studied, and for this purpose net fecundity and mortality are usually sufficient. Game and fish management, however, may require prediction of population changes to an accuracy of 10 to 20 percent, and for this purpose complete age-specific mortality and fecundity schedules are required. Finally, the human demographer needs to project human population sizes to better than 1 percent accuracy, and to do so needs fecundity and mortality figures by age, sex, socioeconomic class, education, geographical location, and so on. (Lewontin 1988, pp. 7–8)

Lewontin's principle is implicit in a report by Ricqlès and Bolt. These workers inquire into the origins and nature of "perhaps the most important captorhinid [prehistoric reptile] character," a feature whose evolutionary history they regard as a prerequisite to proper taxonomic use: jaw growth and tooth replacement (1983, p. 7).[3] Their predictions, though far from precise, are nevertheless legitimate, since their criteria are well-defined and consistently applied. On the basis of large, clearly visible anatomical features, these scientists create a preliminary model of jaw growth and tooth replacement. For verification, they check their predictions against effects of jaw growth and tooth replacement not visible to the naked eye. It makes sense that the dynamics of these processes will leave traces in surviving fossils, and that, if the model is correct, effects visible to the naked eye will be consistent with effects visible at between two and six magnifications.

Among the papers in evolutionary biology I have examined, falsifiability is mentioned almost as much as prediction. Falsifiability is prediction turned on its head; it asserts that a theory can be undermined by its repeated failure to predict. This notion has Sir Karl Popper as its best-known exponent (1965, 1968; see also Medawar 1984). For Popper, the possibility of falsifiability is at the root of all good science, all of which is open to being *"refuted by experience."* In other words, it is the essence of science that we can refute theory with contravening facts:

> It is possible by means of purely deductive inferences (with the help of the *modus tollens* of classical logic) to argue from the truth of singular statements to the falsity of universal statements [in the texts examined, from "observations" to "hypotheses" or "models"]. Such an argument to the falsity of universal statements is the only strictly deductive kind of inference that proceeds, as it were, in the 'inductive direction'; that is, from singular to universal statements.

Popper recognizes that theories under empirical attack can always be saved "by introducing *ad hoc* an auxiliary hypothesis, or by changing *ad hoc* a definition." His solution is "the empirical method" with "its manner of exposing to falsification, in every conceivable way, the system to be tested" (Popper 1968, pp. 41–42).

Two papers in my sample claim an openness to falsifiability. Writing about the spread of chipmunk populations, B. D. Patterson says of a hypothesis integral to his model that it is "apparently refuted by the presence of *E. canipes*, rather than *E. quadrivittatus* [species of chipmunk], in the Gallinas mountains"; according to this model, "small mountain ranges isolated from centers of endemism [areas where species originate] should have been colonized by southwardly advancing taxa," of which *E. quadrivittatus* is one (1982, pp. 393–394). In their paper, Ricqlès and Bolt assert that "evidence for successive occupation by several generations of teeth at the same site would falsify [their] model" of reptilian jaw growth and tooth replacement, a model that implies "that in the MR [multiple-rowed] area a given attachment site was occupied only once" (1983, p. 22). Since the geography of Arizona and New Mexico allows for the apparently anomalous presence of *E. canipes* in the Gallinas mountains, Patterson's hypothesis survives; since successive same-site tooth occupation is not in evidence, the conclusion of Ricqlès and Bolt still rests on firm ground.

The rational reconstruction of the concept of species by means of an analysis of typical papers in evolutionary taxonomy supports the view that science is an inductive progress to reliable knowledge, knowledge always open to, but nevertheless so far resistant to, falsification. The induction from which this knowledge proceeds has observations as its raw material, and these are apparently free from the evolutionary theory that explains them. For example, the collection of sixteen similar birds by Fitzpatrick and his co-workers is a task possible to earlier, evolution-free centuries. It need have been motivated by nothing more than the hunch that the birds might form a new group. Of course, once the group has actually been segregated, it can be defined taxonomically; and, once it has been so defined, it can be redefined according to evolutionary theory. Evolutionary theory can then be used to explain and predict events and entities in the natural world and to retrodict these in the fossil record.

The Rhetorical Reconstruction of Species: Creation

Although Perelman and Olbrechts-Tyteca's *The New Rhetoric* does not contain any extended discussion of argument in the natural sciences—no examples from these sciences appear—there is no intent to exclude the natural sciences from the sphere of rhetoric. The authors believe that scientists address their very specialized audiences, audiences that are commonly very small indeed, as a "universal audience" in the sense "that everyone with the same training, qualifications, and information would reach the same conclusions" (1971, p. 34; compare Johnstone 1978, p. 91). By means of the universal audience, then, the natural sciences come within the sphere of rhetoric: the purposes of a rhetorical reconstruction would be served if we were to "characterize . . . the image [the evolutionary taxonomist] holds of the universal audience he is trying to win over to his view" (1971, p. 33). In order to characterize this image, we must find rhetorical equivalents for the stages of the rational reconstruction just presented: we must show how, *sub specie rhetoricae,* scientists create an ontology that persuades their fellows, an ontology in which plants and animals have been brought to life, raised to membership in a taxonomical group, and made to illustrate and generate evolutionary theory.

Potential Species

The first stage, the creation of the potential species, is best elucidated by the New Rhetorical concept of "presence." Through presence, writers place "certain elements" in their discourses, those on which they "[wish] to center attention," in "the foreground of the [reader's] consciousness." Initially, therefore, presence is "a psychological phenomenon"; that on which the mind and senses dwell "is, by that very circumstance, overestimated" (Perelman and Olbrechts-Tyteca 1971, pp. 142, 116–117; compare Gadamer 1975, p. 103).

This reading of presence is consistent with Gestalt principles. According to these, sensations organize themselves into wholes, or gestalten, certain combinations of which seem automatically to be foregrounded; these combinations we see as having shape and substance, outlined against a shapeless and relatively insubstantial background. Gestalten are clearly manifested to sight and hearing, less clearly to the other senses. Attributively, they can refer not only to things experienced, but also to collections of thoughts and ideas (Köhler 1947). On this reading, presence becomes a special case of perception.

Presence so conceived is easily subject to manipulation. In an ordinary way, we see an object that looks like a pencil; later, we grasp it and write with it, confirming our initial impression. But this simple conformity between our perceptions and the world, a conformity that encourages a naive realism, can be dramatically realigned: optical illusions and camouflage are well-known instances. Less dramatically, it is always possible to "adopt a special attitude with regard to the field so that some of its contents are emphasized while others are more or less suppressed" (Köhler 1947, p. 99). In visual fields, for example, manipulation makes both normal perspective and *trompe-l'oeil* possible.

In the description and depiction of the potential species of evolutionary taxonomy, presence is created by two devices: overdescription and multiple sensory perspectives. The first, overdescription, is the characterization of sense objects in detail far beyond a reader's ordinary expectations. In science, the rhetorical effect of presence routinely generated by overdescription has been analyzed by Steven Shapin; he demonstrates that Robert Boyle's resort to overdescription in his experimental papers was a deliberate attempt to create "virtual witnessing," the use of circumstantial detail as a surrogate for the actual experience of verification through experimentation. In evolutionary taxonomy, overdescription serves an analogous purpose: on their own, words and pictures bring potential species to life.

In pursuit of their ontological goal, evolutionary taxonomists also make use of multiple sensory perspectives. Fitzpatrick and his co-workers, for example, are no purists, looking for essential differences; all descriptive features are mustered that work to increase and individuate presence. This multiplicity of perspectives combined with the extraordinary detail with which each is described creates presence so effectively because it mimics the methods by which we confirm our everyday impressions of the reality of ordinary objects. We supplement visual data with more visual data from different angles and distances; the data of one sense with the data of another; qualitative sense data with mensuration. Thus, the presence of a species increases as the paper progresses; the "whole field of consciousness" becomes filled with this creature "so as to isolate it, as it were, from the [reader's] overall mentality" (Perelman and Olbrechts-Tyteca 1971, p. 118).

In scientific as distinct from literary prose, the resources of language from which presence is created are decidedly limited. Because scientific prose is designed to create the impression that its language refers unproblematically to a real world existing independently of any perceiving subject, it generally excludes the subjective dimension of description, the use of emotion-charged words or irony. For the same reason, scientific prose generally excepts any device that shifts the reader's attention from the world that language creates to language itself as a resource for creating worlds. These restrictions account for the unavailability of many of the techniques of presence discussed in *The New Rhetoric:* onomatopoeia is just one example of such an exclusion.

But language so constrained is hardly free from rhetorical effect; as Barthes cogently observes: "the zero degree . . . *is a significant absence* . . . the absence of rhetorical signifiers constitutes in its turn

a stylistic signifier"; "denotation is . . . the *last* of the connotations
. . . the superior myth by which the text pretends to return . . . to
the nature of language, to language as nature" (his emphasis; 1968,
pp. 77–78; 1974, p. 9). Since the creation of presence in science is
limited to those devices by which language may be said to refer
unproblematically to a real world, it seems fair to give that sort of
presence a name: to call it referential presence.

Taxonomic Species

We can account for referential presence *sub specie rhetoricae;* but can
we account for its taxonomic transformation? We can, if we can show
how statistical inference, naming, and artistic rendering create the
persuasive structures that transform referential presence into taxonomical species.

In evolutionary taxonomy, statistical inference is designed to convince readers that speciation partakes of the certainty generally attributed to a branch of mathematics. There is no question of the importance to evolutionary taxonomy of this frequently used, routinely
foregrounded set of techniques. But it is far from clear that statistical
inference is the actual basis for the establishment of new taxa. The
validity of such inferences depends on the efficacy of sampling procedures: generalizations from a sample can be made only on the basis
of random selection from a defined population. In practice, limitations of time and funding force reasonable compromises. But Fitzpatrick and his co-workers define *Heliangelus regalis* by means of sixteen
birds! How can they assert that these are representative? This practice
of unsystematic sampling, which seems to be the taxonomic norm
(Sokal and Crovello 1984, pp. 544, 558), provides a clue to the
rhetorical nature of the taxonomic transformation.

In truth, taxonomic speciation depends less on statistical inference
than on a fundamental presupposition about the existence of species
in the order of nature. These species must necessarily be imagined
as "discrete sets with appreciable distances between them." Taxonomic speciation absolutely "depends upon the possibility of grouping
data in clusters with empty space between them" (Kuhn 1977, p. 312
and note). This continuity within, and discontinuity between, species
is perfectly in accord with Gestalt psychology: between gestalten there

is "a 'dead' interval which corresponds . . . to the mere extension or ground outside a visual shape" (Köhler 1947, p. 111). But this presupposition is not in accord with the conviction that species are established solely on the basis of statistically inferred regularities.[4]

A striking instance of the operation of the presupposition concerning the essential nature of species, culled from the papers I examined, concerns *Trachelyichthys decaradiatus*—a new species of a new genus of catfish identified and described by Greenfield and Glodek on the basis of a single specimen! Although statistical inference is invoked by the researchers, another, antithetical procedure is actually used. Greenfield and Glodek simply assume synecdoche: a single creature automatically evokes the order of nature as a taxonomic network. In light of such imaginative leaps, one may question the legitimacy with which the precision demonstrated for specimens is routinely transferred to the species.

Artistic rendering also confers taxonomic identity. In Fitzpatrick, Willard, and Terborgh on hummingbirds, the full-color portrait of *regalis* in situ may strike the naive viewer as a candid glimpse. But this picture is clearly posed: it presents these creatures in a manner designed to display not just any characters, but only those that best distinguish them from their fellows. In the case of the line drawings reproduced in Figure 1, the purposes of taxonomic identification are served by another form of visual rhetoric: tendentious simplification.

Finally, naming confers taxonomic identity. As Victor Turner has demonstrated for the Ndembu (1981, pp. 85–86, 180) and David Hull has shown for taxonomy (1984, p. 638), naming is a carefully guarded cultural resource whose purpose is to bestow identity: taxonomic names bestow taxonomic identity. Although taxonomic naming persuades us to credit the scientist with discovery, on the contrary, "wherever language and men exist, there is not only a freedom from the pressure of the world, but this freedom from the habitat is also freedom in relation to the names that we give things, as stated in the profound account in Genesis, according to which Adam received from God the authority to name creatures" (Gadamer 1975, p. 402). The verbal portrait of the hummingbird consists of scattered islands of detail. Naming alone resolves the paradox by means of which this scattered detail becomes a new creature, one for whom a taxonomic space already exists (Barthes 1974, pp. 60–63, 94–95, 209–210).

Evolutionary Species

The evolutionary species is part of at least two networks of meaning, two versions of evolutionary theory. In the first, or strong, version, the species is an integral part of a formulation that aspires to make quantitatively precise forecasts of events in space-time, events directly inferred from mathematically expressed physical laws. In the second, or weak, version, this same concept is an integral part of a formulation that brings together a large and otherwise disparate number of phenomena in the natural world under a single conceptual umbrella.

Darwin himself is the source of both versions of evolutionary theory. In *The Origin,* taking his cue from the explanatory model of classical physics, he says in support of the strong version: "Throw up a handful of feathers, and all must fall to the ground according to definite laws; but how simple is the problem where each shall fall compared to that of the action and reaction of the innumerable plants and animals which have determined, in the course of centuries, the proportional numbers and kinds of trees now growing on the old Indian ruins!" (1872, p. 86). In his letters, writing in the same vein, he compares his theory with gravitation and with the wave theory of light (1972, I, 150; 1959, II, 80, 83–84).

At other times, however, Darwin seems to have explanatory goals for evolutionary theory that are far less rigorous. In his letters, he also says: "An hypothesis is *developed* into a theory solely by explaining an ample lot of facts"; it "[connects these facts] under an intelligible point of view." In a letter to Lyell, though venturing a weak prediction concerning the fossil record, Darwin specifically abjures strong prediction, which he calls "prophecy" (1959, II, 80; p. 210; pp. 9–10). Indeed, as a whole, *The Origin of Species* exemplifies the weak version of evolutionary theory. But to adopt this version is also to abjure falsifiability as a criterion for the adequacy of theoretical formulations.

Are contemporary taxonomists strong or weak theorists? In modern taxonomy, as in Darwin, the weak version of evolutionary theory is predominant. In their work on a new species of humming-bird, Fitzpatrick and his collaborators are typical in their use of convergence to explain an anomalous character; in so doing, they are using evolutionary theory to connect "an ample lot of facts . . . under an intelligible point of view." In the papers I examined, however, Ricqlès and Bolt and B. D. Patterson clearly underwrite falsifiability as the chief criterion for the theoretical adequacy of their formulations. In so doing, they seem to accept the evolutionary species as part of a strong theory. But appearances are deceptive. In his analysis of falsifiability, Popper insists that the element of risk be maximized: "an attempt to solve an interesting problem by a bold conjecture, *even (and especially) if it soon turns out to be false*" (1965, p. 231; his emphasis). Although the statements of Ricqlès and Bolt look Popperian at first glance, careful scrutiny reveals the hyperbolic nature of their claims. In these papers, the theories of their authors are not seriously at risk; *a fortiori* evolutionary theory is never seriously at risk.

In truth, Ricqlès, Bolt, and Patterson are taking advantage of the impressiveness that an openness to falsification confers without taking the risks that such openness should entail.[5] What Ricqlès and Bolt identify as falsifiable is not even central to their argument; it is only their preliminary model, a model which is no more than an artifact of their method. They elected to build this model from large, clearly visible anatomical features, and to correct it by means of anatomical features not visible to the naked eye. As they readily admit, they could have built their final model directly and entirely from submacroscopic features. If they had, potential falsifiability would have disappeared, since "[submacroscopic] evidence for successive occupation by several generations of teeth at the same site" would have been clearly absent from the beginning (Ricqlès and Bolt 1983, p. 22).

Patterson is no truer to Popper than are Ricqlès and Bolt. Because his model really is endangered by some awkward evidence, it is technically falsifiable. But, to save it, Patterson resorts to an *ad hoc* hypothesis, expressly forbidden by Sir Karl: "An elevated corridor . . . extends from the White Mountains north to the Gallinas Range. Chipmunks *evidently* [my emphasis] were able to move through this corridor to colonize the Gallinas Range before it could be reached by *E. quadrivittatus*" (1982, p. 395). One can only conclude that falsifiability is invoked in both these instances to give a mistaken impression of the strength of particular taxonomic claims.[6]

Despite differences, however, the weak version of evolutionary theory is similar in important ways to the strong. In both, natural selection creates the phylogenies, the lines of descent we observe in nature and in the fossil record; in both, evolution is a process theory and a theory of descent. It is these shared theoretical resources that allow taxonomists to use either version as a completely unquestioned, utterly reliable source of intellectual underpinning, an ideational account on which they can reliably draw to explain and extend their observations. Since evolutionary theory founds evolutionary biology, by definition it has axiomatic status: in every paper, its appropriateness in species identification can be taken as settled.

In fact, it is far from settled. When Sokal and Crovello assert that taxonomy can proceed very well without the evolutionary species, "a theoretical ideal to which existing situations are forced to fit as closely as possible" (1984, p. 560), they are not alone, nor is their alternate approach—phenetics—without adherents. And phenetics is not the only alternative. An advocate of "transformed cladistics" believes "that much of today's explanation of nature, in terms of neo-Darwinism or the synthetic theory, may be empty rhetoric" (C. Patterson 1982, p. 119).

At the heart of this problem is a paradox inherent in any taxonomic affirmation of evolutionary theory. If the theory is right, species cannot be natural kinds, entities with atemporal identities, like quarks or electrons: full knowledge of intermediary varieties would demonstrate their wholly historical nature. Concerning Ernst Mayr's classic *Systematics and the Origin of Species,* Niles Eldredge says: "His treatment of the entire problem implies a recognition that if one goes too far in embracing the principles of natural selection and adaptation as the be-all of evolution, the systematist [taxonomist] is left with nothing to explain" (1982, p. xix). In other words, any version of evolutionary theory eventually leads to the disappearance of the species as a legitimate natural kind. The evolutionary species is a rhetorical construct, an oxymoron created only by avoiding the full implications of the theory on which its existence apparently depends.

Rival Reconstructions

An Epistemological Problem. Joseph Gusfield makes a strong claim for the adjunct status of rhetorical analysis: "The rhetorical component *seems* to be unavoidable if the work is to have a theoretical . . . relevance. Thus an analysis of a scientific work *should* . . . include its rhetorical as well as its empirical component" (1976, p. 31; his emphasis). Gusfield's claim depends on a distinction that privileges science at the expense of rhetoric. This view has a clear and venerable source: Aristotle asserts that rhetoric is not a field because it lacks a proper subject matter; furthermore, "in so far as anyone tries to construct either dialectic or rhetoric not as a knack but as a science, he will unconsciously destroy their nature, by passing over, in his attempt to reconstruct them, into sciences of definite subject-matters, and not of mere arguments" (Ross 1971, p. 272). In issuing this warning, Aristotle shares with Gusfield a view of the epistemological superiority of the nomological over the hermeneutic disciplines.

When this view is set aside, the way is open to acknowledge the disciplinary status of rhetoric. Rhetoric has as its subject matter the persuasive structures of all fields, itself included. As the study of such structures in the sciences, rhetoric of science presupposes that all components of disciplinary discourse are within its explanatory compass: there is no empirical or theoretical core, no essential science that reveals itself all the more clearly after the rhetorically analyzed components have been set aside. As a discipline, rhetoric may be expected to behave differently from classical physics; one does not expect laws of rhetoric or rhetorical predictions. But the disciplinary claim is fundamental: without disciplinary status, rhetoric of science is not a field but a bundle of techniques, an adjunct to fields.

When we ignore Aristotelian biases, we can easily set rhetoric beside such hermeneutical disciplines as history, biblical commentary, and literary criticism; indeed, its disciplinary status is far more ancient: the case can be made that rhetoric was the first organized body of hermeneutical study, the *ur-Geisteswissenschaft*. Throughout its long career, rhetoric has been overshadowed first by philosophy, then by science, both of which limited its scope and trivialized its usefulness. But such limits and such trivialization are not inherent in the nature of rhetorical theory; rather, they are the result of a continuing need in Western civilization to open up a space for the possibility of certain knowledge.

The disciplinary claim for rhetoric of science has as its source a fundamental federalism about the domains of knowledge. In accordance with this federalism, rhetoric is one discipline among many in a joint enterprise, a confederation of equally sovereign intellectual states. A central authority is deliberately absent; claims to such authority—the traditional claims of philosophy and theology—are wholly without merit. No discipline is, or can be, privileged over another. This chapter exemplifies the absence of such privilege: it reconstructs evolutionary taxonomy in two equally valid ways.

According to its rational reconstruction, evolutionary taxonomy is a discipline at whose center lies the evolutionary species; in fact, belief in the reality of such species is the *sine qua non* of being an evolutionary taxonomist. This rational reconstruction of science is both justificatory of, and coextensive with, science itself. For rhetoricians of science, on the other hand, the reality of the evolutionary species is essentially textual. Rhetoricians insist that scientists "*establish the real. The real* is as much a hypothetical construct as is the universal audience" (Karon 1976, p. 103).

Rational and rhetorical reconstruction differ in their fundamental motivation. Rational reconstruction is clearly subsumed under "the cognitive interest in technical control over objectified processes" (Habermas 1971, p. 309; see also p. 212). Rhetorical reconstruction, on the other hand, is a manifestation of the practical interest "oriented toward mutual understanding in the conduct of life." This interest is "directed toward the transcendental structure of various actual forms of life, within each of which reality is interpreted according to a specific grammar of world-apprehension and of action" (Habermas 1971, pp. 311, 195). The structures of persuasion of evolutionary taxonomy are part of such a grammar.

Both reconstructions of evolutionary taxonomy are equally legitimate; each, though incomplete in the larger view, is complete in itself. But an autonomy so radical creates an epistemological problem. Each interest, the technical and the practical, has its appropriate epistemology, its bundles of pertinent analytical techniques. By way of contrast, the theoretical vantage of this chapter lacks legitimation. By what right has evolutionary taxonomy been reconstructed, its reconstructions compared, and their ontological equipollence inferred?

An Epistemological Solution. Traditionally, transdisciplinary objectivity depended on our ability to "rise above" the narrow views of the "petty" kingdoms of disciplinary knowledge; to move outside their "confining" boundaries; to transcend their "parochial" concerns. Such objectivity licensed the class of statements at this chapter's theoretical core. But my analysis of evolutionary taxonomy seems to confirm Wittgenstein's well-established view that objectivity consists in living by the laws of one intellectual state or another (in Wittgenstein's terminology, one form of life or another). Outside of these there is apparently nothing, no real world we can use for our criterion.

So strict an interpretation of Wittgenstein has a certain plausibility. In particular sciences, there is the common experience that central concepts increase in ontological status, in objectivity, wholly within a single domain of interest, the technical. Did not Einstein's papers on Brownian movement put to rest any doubt about atomic reality? Did not Watson and Crick's discovery of the structure of DNA finally give the functional description of the heredity process a firm physical sense? But additional reference to the history of science reveals the incompleteness of this view. Galileo's battle against geocentricity, Young's struggle against the particle theory of light, Darwin's war against the argument from design: in each of these instances a central concept, a scientific "fact"—geocentricity, particle theory, design— was revealed as a presupposition rooted not only in the technical but also in the practical interest, in tradition and in current, widespread use.

A strict interpretation of Wittgenstein, then, leads to a misunderstanding of science as a wholly rational pursuit. Indeed, the rational and the rhetorical are far from opposites; their opposition, so deeply a part of our intellectual heritage, is alleged, not proven. To undermine the Sophistic tenet that knowledge was rhetorically constituted, Plato and Aristotle drew a firm line where none had existed. Above *doxa*, a knowledge no more privileged than its knowers, they raised *episteme*, true knowledge: the accurate depiction of an independent reality.

Without this prestigious endorsement, an endorsement whose influence is still strong within Western thought, would we not see similarity where we now see only disparity? Where would definition by family resemblance be without the *topoi* of genus and difference? Need classification be independent of the *topos* of comparison? Is there an *essential* difference between scientific and rhetorical description? Between scientific and rhetorical definition? Is not the application of evolutionary theory a systematic use of the *topos* of antecedent and consequence? Is not the conflict between the rational and the rhetorical a civil war?

To negotiate a truce in this conflict, we need to draw on a sense of objectivity and rationality broader than that of any discipline. This broader sense has its source in the third, and most important, of Habermas's three human interests, the emancipatory, an interest one of whose purposes is to "[destroy] the illusion of objectivism [cultivated by the sciences] through demonstrating what it conceals: the connection of knowledge and interest" (Habermas 1971, pp. 316–317). The vehicle of this emancipatory interest is critique, a discourse on discourse, dialectic in mode, ironic in manner. It is by means of critique that we reconstruct evolutionary taxonomy in two very different ways, and compare these reconstructions.

It is with critique that I would identify a broader, transdisciplinary sense of rhetoric, a sense different from that exemplified earlier by the application of the conceptual machinery of Aristotle's *Rhetoric* to the texts of evolutionary taxonomy. In this sense, rhetoric is not a discipline but a perspective whose essential character is reflexive and ironic.[7]

In an apt political metaphor, Kenneth Burke captures this perspectival sense of rhetoric: "Insofar as terms are thus encouraged to participate in an orderly parliamentary development, the dialectic of this participation produces (in the observer who considers the whole from the standpoint of the participation of all the terms rather than from the standpoint of any one participant) a 'resultant certainty' of a different quality, necessarily ironic, since it requires that all the sub-certainties be considered as neither true nor false, but *contributory*" (1962, p. 513). This seems exactly right. From the point of view of rhetoric as critique, the rationality of science consists in the continuing dialectic among its legitimate reconstructions, each the surrogate for the informed assent of an interpretive community; analogously, the objectivity of science is constituted by some configuration of these reconstructions. Such configurations must be viewed ironically; only if irony is presupposed do they avoid the charge of inconsistency.

Rhetoric, then, is both a discipline and a perspective from which disciplines can be viewed. As a discipline, it has a hermeneutic task, and generates knowledge; as a perspective, it has a critical, emancipatory task, and generates new points of view. The central goal of this chapter is not hermeneutic, but critical and emancipatory: I have elaborated, illustrated, and refined a new perspective. I hope I have shown that a sharp distinction between rhetoric and rationality is untenable—that such a distinction entails an unduly narrow view of both. Finally, I have tried to show that the objectivity of evolutionary taxonomy depends as much on rhetoric as it does on rationality.

CHAPTER 4

The Tale of DNA

Despite its formidable title, "A Structure for Deoxyribose Nucleic Acid" by J. D. Watson and F. H. C. Crick has not daunted rhetorical critics (see Bazerman 1981; Carlisle c. 1983; Halloran 1980; Limon 1986). Rightly so; the persuasive effect of this short communication was enormous. Within a dozen years, its findings permanently altered the character of biological research. In this chapter I do not intend merely to rehearse, to deepen, or to extend the claim of such critics that Watson and Crick use persuasive devices to convince scientists of the correctness of their structure; rather, I want to support a more radical claim: that the sense that a molecule of this structure exists at all, the sense of its reality, is an effect only of words, numbers, and pictures judiciously used with persuasive intent.

This more radical case is often made for science in general, but without firm evidence; or it is made for the "soft" sciences, or for social "science," not for a "hard" science like stereochemistry (see Brummett 1976; Kelso 1980; Douglas 1971; Overington 1977). But the radical case is perfectly general. We make ontological commitments all the time; our everyday speech is full of assertions of what is. These assertions depend for their credibility on their fit with previous ontological commitments. The ontological commitments of science are distinguished from those of everyday life only by the greater rigor of the criteria applied to them, criteria derived from mathematics and logic: stereochemistry is unlike everyday life in the precision of fit required for rational commitment. But exactness of fit is a strategy for increasing the precision, not the degree of commitment: mathematics and logic are formal disciplines that can add nothing to our sense of what is (Quine 1961).

To help establish this absence of ontological privilege for scientific knowledge, and to assert the epistemological pluralism it implies, it is necessary to show that science shares essential characteristics with other routes to knowledge, routes as different as autobiography, history, and literary criticism. In the case of the discovery of the structure of DNA we are fortunate to have two versions to compare, versions apparently dissimilar in every way: the scientific account in Watson's first communication, written with Crick, and Watson's autobiographical account in *The Double Helix*. In this chapter I will show that both accounts can be analyzed in a way that legitimately emphasizes their considerable common ground. The scientific communication succeeds in persuading peers that the double helix exists; the autobiography succeeds in convincing those same peers that Watson's view of scientific practice is essentially correct. In both cases the warrant for belief is the same: the fit between the new view presented and a preferred view of the constitution of reality.

The Double Helix as Narrative

In 1953 two obscure scientists, James Watson and Francis Crick, published a paper that, by the general agreement of their peers, revolutionized twentieth-century biology. In it, they suggested a structure for DNA that made excellent sense of the available evidence. DNA had for some time been suspected of holding the key to heredity, and Watson and Crick's proposed structure—two intertwined helices or "spirals"—also hinted strongly at the way reproduction, and thus hereditary transmission, might take place at the molecular level. Untwisted and separated, each partner of a pair of helices might act as a template or mold from which new partners would be formed.

In 1967 Watson published an autobiographical memoir of his joint discovery, *The Double Helix*. Although some critics voiced objections to the fierce competitiveness he seemed to endorse, an abundance of authoritative testimony assures us that Watson's impressions are not essentially idiosyncratic, that his is an authentic view of the world of contemporary science. André Lwoff speaks of the work's conformity to truth: it is "a fascinating book. For the first time all the steps and circumstances of a major contribution to science are described with precision and accuracy" (1968, p. 137). Jeremy Bernstein speaks of its human qualities: it is "a unique work about science. It concerns a great discovery and it is written by one of the discoverers soon enough after the event so that all the human details that make a work of science so much like a work of art have not been forgotten" (1968, p. 182). Jacob Bronowski speaks of its essence: it "communicates the spirit of science as no formal account has ever done" (1968, p. 382).

This chorus of praise remains surprising. Though Watson's account has considerable narrative presence,[1] his story seems clearly unflattering to scientists. According to *The Double Helix,* the discovery of the structure of DNA was made by two bunglers, of whom Watson is the chief. As a graduate student at Indiana University, he hopes to solve the problem of the gene without "learning any chemistry." His lack of chemical training is the result of total incompetence: "After I used a bunsen burner to warm up some benzene, I was relieved from any further true chemistry." He is similarly inept at crystallography and ignorant of the mathematics essential to that discipline. Crystallographers have made great strides in solving molecular structures; it is, in fact, crystallographic evidence that will eventually confirm the Watson-Crick structure of DNA. Yet Watson is "ignorant of Bragg's law, the most basic of all crystallographic ideas." Crick, more adept than he at mathematics, explains something about crystallography, but "the mathematics eluded me." Although he sees the importance of crystallographic work, Watson is "unable to understand large sections of [a] classic paper" on the subject (1966, pp. 21, 41, 77, 111).

This bungler and his collaborator have two main rivals: Rosalind Franklin in England and Linus Pauling in the United States. Called Rosy behind her back, Franklin fairly bristles with hostility. She has an "acid smile"; when she speaks, there is "not a trace of warmth or frivolity in her words" (1966, pp. 148, 68). Afraid of her "sharp retort," afraid "to be told by a woman to refrain from venturing an opinion on a subject for which [they] were not trained," male scientists avoid her. Her hostility is more than merely verbal. In one confrontation, Watson is afraid "that in her hot anger she might strike me." This fear is confirmed when he learns from Wilkins, Franklin's laboratory supervisor, that "some months earlier she had made a similar lunge toward him." Throughout, it remains clear that "the best home for a feminist was in another person's lab" (1966, pp. 70, 166, 167, 20). In contrast to Franklin, Linus Pauling, Watson's more important rival, is treated with great respect. He is "the greatest of all chemists" and "the world authority on the structural chemistry of ions." His book, *The Nature of the Chemical Bond,* is a "masterpiece" (1966, pp. 18, 80, 101).

Despite their ignorance and their awareness of these formidable rivals, despite the fact that "the odds appear against" such "dark horses," Watson and Crick persist in "racing [Pauling] for the Nobel Prize" (1966, pp. 163, 130, 184), while continuing foolishly to misunderstand Franklin's work. Following this stumbling course, they nevertheless create a structure and invite some fellow scientists—including Franklin—to view and admire it. View it she does, but objects to the model on sound scientific grounds, revealing "the embarrassing fact [that Watson's] recollection of the water content of Rosy's DNA samples could not be right. The awkward truth became apparent that the correct DNA model must contain at least ten times more water than was found in our model." It becomes clear that the "fifty-mile excursion" of Franklin, Gosling, her assistant, and Wilkins has been taken in the interest of the "adolescent blather" of two scientific clowns (1966, pp. 94–95). Partly as a result of this fiasco, Watson and Crick are forbidden by Bragg to work on DNA.

Nevertheless, they persist and triumph. Franklin takes a wrong path, arguing against helices, and Pauling stumbles badly. He publishes a structure for DNA, but his paper contains an elementary chemical blunder, apparently the product of haste. Watson estimates that he has six weeks to solve the structure; amazingly, within this period of time the solution is in his hands. True, without Chargaff's base ratios, without Donahue's advice about the keto forms, without the work of Franklin, Gosling, and Wilkins, without the inside information provided by Peter Pauling, Linus's son, without the somewhat bemused tolerance of Sir Lawrence Bragg, the great crystallographer, now administrator of the Cavendish, the laboratory to which they were attached, and, finally, without each other, Watson and Crick would never have succeeded. But when success came, it was certainly sweet and the triumph was for each of them to savor:

> Chargaff's rules suddenly stood out as a consequence of the double-helical structure for DNA. Even more exciting, this type of double helix suggested a replication scheme much more satisfactory than my briefly considered like-with-like pairing. Always pairing adenine with thymine and guanine with cytosine meant that the base sequences of the two intertwined chains were complementary to each other. Given the base

sequence of one chain, that of its partner was automatically determined. Conceptually, it was thus very easy to visualize how a single chain could be the template for the synthesis of a chain with a complementary sequence.

The results were "almost unbelievable." They had made a discovery that would "revolutionize biology"; Crick felt "the pitch of his excitement . . . rising each day." Those around the two young scientists were "participating in perhaps the most famous event in biology since Darwin's book" (1966, pp. 196, 198, 199, 214, 220–222).

To the legitimacy of this jubilation Linus Pauling and Rosalind Franklin give their clear warrant. Franklin simply and unequivocally agrees. Pauling, visiting the Cavendish on the way to a Solvay conference in Belgium, unhesitatingly capitulates: "All the right cards were in our hands and so, gracefully, he gave his opinion that we had the answer" (1966, p. 222).

The Double Helix as Rhetoric

In *The Double Helix* Watson generally gets his details right; indeed, he exercises a vigilance over detail exacting enough to win praise from one of his harshest critics.[2] But his insight into other people's inner lives is often wide of the mark; he shows a defect in understanding clearly supported by the testimony of Wilkins and Chargaff concerning their own motives. Moreover, there is considerable evidence that the portraits of Franklin, Linus Pauling, Crick, and Watson himself are seriously distorted.

Rosalind Franklin's acceptance of his model "amazed" Watson (1966, p. 210), but there is no reason for this amazement, since the facts indicate that he and Franklin were thinking along compatible lines.[3] Watson's sarcastic reaction to Pauling's acceptance is just as odd. At Solvay the famous chemist generously announced that it was "very likely that the Watson-Crick structure is essentially correct" and that "the formulation of their structure . . . may turn out to be the greatest development in the field of molecular genetics in recent years" (1952, p. 113). Moreover, although Pauling wanted badly to solve the problem of DNA structure, he was not a mere six weeks away from the solution. In retrospect, he stated that "if Watson and Crick had not carried on their persistent effort . . . the discovery of

the double helix . . . might well have been delayed for several years"
(1976, p. 771).[4]

Thus Watson's fears seem self-induced and his enemies largely of
his own making. They were merely good, ambitious scientists, as was
he, despite his constant picture of himself as a bumbler and a bungler.
His training was excellent. His mentors, Luria and Delbrück, both
won the Nobel Prize; his laboratory, the Cavendish, presided over by
a Nobel laureate, was one of the finest in England. After the Prize,
Watson chose a largely administrative career, but his world-famous
laboratories at Cold Spring Harbor have "the best students, the best
postdocs and do the best work" (quoted by Zuckerman 1977, p. 235).
Crick, of course, had a distinguished scientific career subsequent to
the discovery; he is now Sir Francis Crick. At the time *The Double
Helix* was published, there was a rumor that he so strongly dissented
from Watson's portrait of him that he might sue his former collaborator. Asked twenty-one years after the discovery to reminisce about
it, Crick fantasizes an autobiography he might write, paralleling *The
Double Helix:* it would be entitled *The Loose Screw,* and its opening
sentence would be a wicked parody of Watson's own opening (1974,
p. 768).

Watson's autobiographical distortions have a definite pattern: he
consistently casts himself in the role of the youngest son of the fairy
tales. In "The Queen Bee," for instance, Blockhead (Dummling), the
youngest of three brothers, wins the hand of a beautiful princess by
accomplishing three tasks, the first of which involves the collection
of a thousand pearls strewn about in a forest. Although the two older
brothers jeer at the youngest, they fail at the task and are turned
into stone. Dummling succeeds, but only with the help of a society
of ants whose lives he had saved when they were at the mercy of his
siblings. He accomplishes the remaining tasks in a similar manner,
with the help of the ducks and the bees. As a result, he "married the
youngest and most attractive daughter and became king after her
father's death" (Grimm and Grimm 1980, p. 242; my translation).

According to the motif embodied in this tale—and in "The Golden
Bird" and "The Golden Goose," among others—life is a competitive
struggle whose highest rewards are unavailable without the merited
help of others, usually animal others; with this help the hero's success
is assured, despite his reputation for feeblemindedness. In short,
these tales are all variants of one fundamental underlying narrative

structure. Speaking of the folk tales of which these stories are variants, Propp says: "Like any living thing, the tale can generate only forms that resemble itself. If any cell of a tale organism becomes a small tale within a larger one, it is built . . . according to the same rules as any fairy tale." These rules are not a hindrance to creativity on the part of tellers of tales; they merely "demarcate those areas in which the folk narrator never creates, and areas in which he creates more or less freely" (1984, pp. 78, p. 112).[5]

It is no accident that this underlying narrative structure repeats itself from tale to tale, for its attractiveness has deep psychological roots:

> Clearly, this is a child's fantasy of growing up: If only we could recall how we felt when we were small or could imagine how utterly defeated a young child feels when his . . . older siblings temporarily reject him or can obviously do things better than he can, or when adults—worst of all, his parents—seem to make fun of him or belittle him, then we would know why the child often feels like an outcast: "a simpleton." Only exaggerated hopes and fantasies of future achievements can balance the scales so that the child can go on living and striving. (Bettelheim 1977, p. 125)

In this reading, Linus Pauling and Rosalind Franklin are the intimidating grownups or older siblings and Watson is the simpleton: Pauling is Watson's awe, Franklin, his anger. The assistance of Crick, of Wilkins, and of Donahue, which leads, finally, to the intuition allowing Watson to solve the structure, has its parallel in the helping animals; the Nobel Prize is the equivalent of the hand of the beautiful princess and the inheritance of the kingdom.

This reading is not mine *de novo;* it is prefigured by the book's reviewers, especially its scientific reviewers. F.R.S. refers to "Sir Lawrence Bragg, the bad fairy stepmother of the whole tale" (1968, p. 62), and Jacob Bronowski calls the story "a classic fable about the charmed seventh sons, the antiheroes of folklore who stumble from one comic mishap to the next until inevitably they fall into the funniest adventure of all: they guess the magic riddle correctly" (1968, p. 381). Peter Medawar speaks of Watson's "childlike vision" (1968, p. 5), and Robert Sinsheimer, an unfavorable critic, is right on the mark when he says that the book is "filled with the distorted and cruel perceptions of childish insecurity. It is a world of envy and intolerance, a world of scorn and derision" (1968, p. 4; see also Lwoff 1968, p. 137).

Watson's choice of this underlying pattern from fairy tales, and his reliance on its psychological force, are deliberate strategies designed to recreate the state of mind of a young scientist caught in the swirl of his own feelings and the strong currents of an exciting time. He says in his preface: "I have attempted to re-create my first impressions of the relevant events and personalities rather than present an assessment which takes into account the many facts I have learned since the structure was found. Although the latter approach might be more objective, it would fail to convey the spirit of an adventure characterized both by youthful arrogance and by the belief that the truth, once found, would be simple as well as pretty" (1966, p. xi).

That Watson's choice of psychological over literal truth is deliberate is also supported by the book's structure. The central narrative of *The Double Helix* is framed by an untitled prologue set three years after the event and an epilogue set in the writer's present, and the tone these two sections share differs markedly from the rest of the book. The prologue and epilogue are literary devices used by Watson to see his earlier self in perspective. As a younger man, he really did perceive Rosalind Franklin in very unflattering terms; on reflection, in his epilogue, he "[realizes] years too late the struggles that the intelligent woman faces to be accepted by a scientific world which often regards women as mere diversions from serious thinking" (1966, p. 226).[6]

In Richard Lewontin's apt formulation, *The Double Helix* "is an engaging and sometimes exciting book . . . because it speaks to [the] secret dreams [of scientists] in a familiar vocabulary" (1968, p. 2). This is the crucial point. The narrative presence that constitutes the chief ingredient of Watson's literary talent is persuasive to scientists because it is harnessed in the interest of a particular view of life. Many of the book's scientific reviewers responded enthusiastically to Watson's deliberate misreading of his past; they did so because they found compelling the book's underlying narrative pattern, a pattern whose roots reflected a truth deeper than historical accuracy. These scientists saw their world as competitive, a place that was especially contentious when, in their youth, they had their reputations to make. But they had triumphed, just as Watson, as Dummling, did. In other words, scientists found *The Double Helix* persuasive because its view of life so closely matched their own; there was a fit between the view Watson presented and their own preferred view of reality.[7]

A Rhetorical Analysis of Watson and Crick's Paper

In *The Double Helix* Watson mentions reading an article by Linus Pauling, his rival. Pauling's paper describes another helix, the alpha helix, an important structural element in proteins. Watson says of the paper that "it was written with style"; of following papers by Pauling, he says that the "language was dazzling and full of rhetorical tricks" (1966, p. 35). Whatever the truth of Watson's comments, we cannot doubt that his underlying assumption is generally correct: scientists must persuade their fellows that what they say is correct and important. The marketplace of scientific ideas is so crowded that most scientific papers are ignored; most are seldom, if ever, cited and are, by this neglect, "judged to be either trivial or incorrect" (Gilbert 1976, p. 294). In chemistry, for example, only one paper in a hundred is cited more than 10 times, about one in a thousand 66 times or more (Small 1978, p. 330).

Examining the rhetoric of Pauling's alpha helix paper, we find that he is not averse to self-advertisement. In its exordium, he calls his predictions "reliable," his configurations "reasonable." What he finds is "important" and a "discovery" (Pauling, Corey, and Branson 1951, p. 205). In their first paper on DNA, Watson and Crick imitate the spirit of this self-promotion. In their exordium, they "wish to suggest a structure for the salt of deoxyribose nucleic acid (D. N. A.). This structure has novel features of considerable biological interest." It is on these biological implications that their peroration focuses: "It has not escaped our notice that the specific pairing we have postulated immediately suggests a possible copying mechanism for the genetic material" (1953b, p. 737). In the first passage, the bold connotations of "novel" and "considerable" contrast with the mock timidity of "suggest"; in the second, a repetition of this timid verb underlines the litotes of the opening phrase and the hyperbole of the adverb. In short, we are in the presence of irony.[8]

Before we analyze this irony, however, we need to deal with a possible difference in perception: the rhetoric of these papers may not strike *us* as "dazzling." But it is, after all, a matter of degree: in the context of the paper, words like "novel," "considerable," and "immediately" form in fact a connotational network that is highly foregrounded against the absolutely level, denotational surface of the surrounding language: "acidic hydrogen atoms," "negatively charged phosphates" (1953b, p. 737). For those to whom this contrast may seem trivial, a musical analogy may be helpful: "In a style such as Palestrina's, which uses only relatively very static intervals, small differences in interval tension will have great musical import. So much of his melodic motion is stepwise that *any* skip expresses relative tension" (Erickson 1957, p. 37).

To possess an appropriate scientific presence, a molecular model characterized as novel must be described in a manner adequate to that characterization. To establish precision as their criterion for such adequacy, Watson and Crick first survey existing descriptions. Concerning the model of Pauling and Corey, they question crucial details, like van der Waals distances; concerning the model of Fraser, they question the absence of such details. Thus the same criterion allows Watson and Crick to criticize Pauling and Corey's model because its specific features are out of line with existing chemical knowledge, and to dismiss Fraser's model out of hand because it is "rather ill-defined" (1953b, p. 737). By paying ironic tribute to others' descriptive efforts, Watson and Crick simultaneously establish the importance of their problem and create a conceptual space for their solution.

Their own description is openly designed to avoid the *tu quoque;* it is deliberately vulnerable to the same criticisms brought to bear on rival models. Watson and Crick precisely describe the dimensions and orientation in space of each component of their complex structure. For example, they show a sugar residue perpendicular to each helical chain "every 34 A[ngstroms] in the z direction [the third dimension of Cartesian space]" (1953b, p. 737). To make criticism even easier, they also depict their model: a boldly drawn vertical double helix held in place by a horizontal array of laddered base pairs transected by a vertical axis. Directional arrows represent the opposing senses of the helices. In these ways, verbal and visual cues combine to suggest that the model is a molecule, an entity that is perceptible independent of its description and depiction.

But we must remember that the foundation of this description and depiction is not a physical object, one we can see and touch in its three-dimensional actuality; it is, instead, a two-dimensional x-ray diffraction photograph whose third dimension is entirely the product of inference. Watson and Crick's interpretation *in* words and pictures is an interpretation *of* words and pictures *through* words and pictures.

Watson and Crick devote by far the largest portion of their paper to describing their model of the DNA molecule, a static construction made credible by means of the precision of its fit, the sense it makes of previously isolated chemical facts, especially the fact that the ratios of the base pairs consistently approximate unity. But the achievement of this task accomplishes only the lesser of their two persuasive goals. Watson and Crick promised that DNA was not just another moderately complex molecule, however correctly described, but was also "of considerable biological interest." Given the ironic pregnancy of this assertion, it seems odd that the two researchers should spend so little time in its support. Seemingly, we have only the one sentence: "It has not escaped our notice that the specific pairing we have postulated immediately suggests a possible copying mechanism for the genetic material."

The answer to this puzzle lies in the rhetorical function of the adverb "immediately," really an instruction to the reader to re-view the description and depiction of the DNA molecule, to see the dynamic possibilities of an entity hitherto viewed as static. We are asked to perceive a just-described static structure in a new way; to undergo a Gestalt shift. In one sense, "immediately" is a rhetorical exaggeration, a hyperbole; in another sense, it is not. We may not instantly see the dynamic possiblities of the molecule; but once we do, our perception must be immediate. The molecule then fits beautifully into its new, more interesting context, that of Mendelian genetics. It is the fit of the now-dynamic molecule into this second context that fully satisfies the promise of the paper's opening sentence.[9]

I have argued in this chapter that *The Double Helix* and "A Structure for Deoxyribose Nucleic Acid" share not only their subject matter but a persuasive purpose. *The Double Helix* convinced many scientists that it captured the essence of a past parallel to their own; "A Structure" convinced many of these same scientists that the Watson-Crick model was an accurate description of an important molecule. In both the autobiographical account and the scientific paper, style and substance support persuasive intent: we do not have rhetoric on the one hand and science, or autobiography, on the other, but a fusion of the two.

The analysis of "A Structure" parallels that of *The Double Helix,* a parallel that is not accidental. It is true that scientific theories are by design explicit; that, in contrast, Watson's "theory" of his life is implicit. But this difference is not epistemologically significant; it is not an index of an essential difference between two sorts of explanation. It is a mistake to claim that scientific theories generate knowledge, while Watson's "theory" of his life does not. True, the underlying structures of literary works, structures such as allegory, simply increase the coherence of the aesthetic objects of which they are a part; they do not generate knowledge. But the underlying pattern of *The Double Helix* is not such a structure; it is a theory in the strict sense. Because it explains the facts of a life, it generates knowledge, and it is equally susceptible to empirical confirmation or disconfirmation.

In both cases this knowledge goes well beyond the establishment of fact. In the autobiographical case, persuasion is the product of a narrative presence, of facts shaped by adherence to a pattern supposed to underlie a life in science, a "theory" of such a life. In the scientific case, persuasion is the product of a descriptive and processual presence, of facts shaped by an analogous adherence to underlying stereochemical and genetic theories. The point of Watson's narrative is not that these events occurred, but that they had a certain significance; in the scientific case, the structure of the DNA molecule permits, but does not entail, the molecule's startling biological import.

PART II

Style, Arrangement, and Invention in Science

CHAPTER 5

Style in Biological Prose

Speaking of theoretical physics, Einstein says:

> The structure of the system is the work of reason: the empirical contents and their mutual relations must find their representation in the conclusions of the theory. In the possibility of such a representation lie[s] the sole value and justification of the whole system, and especially of the concepts and fundamental principles which underlie it. Apart from that, these latter are free inventions of the human intellect, which cannot be justified either by the nature of that intellect or in any other fashion *a priori*. (1954, p. 272)

Einstein was well aware that "the view I have just outlined of the purely fictitious character of the fundamentals of scientific theory was by no means the prevailing one in the eighteenth and nineteenth centuries" (1954, p. 272). Or the twentieth. Einstein's view entails a distinction between fact and theory, between Snell's Law of refraction and its explanation. Snell's Law is a fact; but the theories that account for it—theories of light and of translucent bodies—are "free inventions of the human intellect," "fictions" in Einstein's sense. If science is about such "fictions," rhetoric has a central role in its analysis; and the proper deployment of the central concepts of rhetoric—style, arrangement, invention—will yield an appropriate intellectual harvest.

I argue in this chapter that one of these central categories, style, is meaningful in the analysis of scientific prose; that the systematic overuse in such prose of certain syntactic options and certain semantic strategies is an integral part of the message science conveys. The features of scientific prose on which I focus are those commonly ascribed to it: examples consistently illustrate the reliance of scientific writing on passive constructions dominated by complex noun phrases.[1] Why is this the case?

Syntax and Semantics in Scientific Prose

A natural language privileges persons; in contrast, the "splinter of ordinary language" (Quine 1966, p. 236) that we call scientific discourse privileges a world from which persons are virtually excluded, a world of physical objects in Quine's sense:

> We might think of a *physical object,* more generally and generously, as simply the whole four-dimensional material content, however sporadic and heterogeneous, of some portion of space-time. Then if such a physical object happens to be fairly firm and coherent internally, but coheres only rather slightly and irregularly with its spatio-temporal surroundings, we are apt to call it a body. Other physical objects may be spoken of more naturally as processes, happenings, events. (Quine 1970, p. 30)

Strawson speaks of the superior "identificatory force" of the subject position in grammar,[2] a position reserved in natural language primarily for speakers and their fellow creatures. In contrast, in science the subject position is reserved primarily for the physical objects Quine posits. By means of this linguistic strategy, science invests such objects with the importance ordinarily bestowed on human beings: *we* are at the causal center of our world; *physical objects* are at the causal center of the world of science.

This linguistic and scientific coincidence means that the progress of science is accompanied by regular linguistic change. At first, anonymous experimental and observational events are named and become scientific terms; then these terms are transformed into the noun phrases of sentences, noun phrases whose syntactic complexity gradually increases. By this means, scientific terms are linked to ever-widening networks of theoretical knowledge.[3]

Roland Barthes's observation concerning realistic fiction can be applied to this process. As these networks widen, the "force of meaning" of scientific terms increases: "The strongest meaning is the one whose systematization includes a large number of elements, to the point where it appears to include everything noteworthy in the world" (1974, p. 154). At this juncture the grand generalizations of science supervene; in these, the complexities of scientific terms, so laboriously compiled, seem suddenly to drop away; actually the new terms of these generalizations *abbreviate* the numerous complex noun phrases that went into their making (cf. Pinch 1985b).

A series of papers on cancer that forms one basis of this study provides numerous examples of this coincidence of scientific and linguistic change (Spector, O'Neal, and Racker 1980a, 1980b; Racker and Spector 1981; Rephaeli, Spector, and Racker 1981; Spector, Pepinski, Vogt, and Racker 1981). In a typical instance, Racker and his associates cease to observe the laboratory event, the blob in the fifth lane of the autoradiogram, and begin to see the physical object: the kinase PK_L. Next, the physical object PK_L begins to appear in the subject position in their sentences. Afterwards, PK_L becomes part of a network of meaning, a causal chain: the kinase phosphorylates, the phosphorylated kinase phosphorylates another kinase. Finally, the nature and activities of all of the phosphorylated kinases are abbreviated in a new noun phrase, a new scientific term: the kinase cascade.

The examination of typical revisions made by journal editors in a set of scientific papers also reveals an overriding need to privilege an ontology of physical objects.[4] In apparent conformity with the standard textbook goal of concision, the editors excise superfluous verbiage:

Original	Revision
"We found no reports in the literature"	"We found no reports"
"It has been recently demonstrated that"	Delete
"Until investigations are carried out of cells exhibiting a wide range of propagating patterns, any further attempts to categorize cell propagating into different types will be of little value."	Delete (Editor's note: "Of course!")

But so necessary to a privileged ontology are the complex noun phrases of scientific prose that they are created by deliberate redundancy. Data already in accompanying tables are repeated in the text:

Original. "After 24 hours transport of ^{32}P into nucleus of RZ cells was high."

Revision. "The level of ^{32}P in nucleus 24 h after applying 0.42 ng MLL-6–^{32}P (rl inc, 77.1 mDo/mmol) to membranes of RZ *Ostitium* cells was high."

By clarifying pronoun references and by placing modifying elements closer to what they modify, the editors apparently adhere to another textbook standard, that of clarity:

Original. "Increasing the concentration of NVM in the enzyme solution did not reduce counts in the nucleus. This despite the fact that the solution used for this treatment had a lower specific activity . . . "

Revision. "Increasing the concentration of NVM in the enzyme solution did not reduce the counts in the nucleus, even when the solution used . . . had a lower specific activity . . . "

Original. "The cylinder was placed standing vertically, with creased edge up, in 100 ml of sterile, deionized water in a sterile 600 ml beaker."

Revision. "The cylinder, with creased edge up, was placed standing vertically, in 100 ml . . . "

But the editors also require the revision of passages that are perfectly clear, revisions whose primary purpose is to make the noun phrases of science the ontological focus of scientific sentences:

I was initially led to this particular study by perusing, during the spring of 1979, the geological bulletin for Lake County, Indiana (PURSE et al., 1975). On page 46 of this bulletin is a reference, with a photograph (taken by GEORGE BARTON, 1975), to a prehistoric camp site along an "east/west road" in the northern part of the county. This photograph (providing geographic coordinates in the legend) illustrates three small "rocks," lying exposed at the site. I was anxious to see if the site could still be found . . .

Requested by a referee to "condense and make this less a folksy narrative," the author did so without demur:

I was initially led to this study by locality data on a prehistoric camp site provided in the geological bulletin for Lake County, Indiana (PURSE et al., 1975). The coordinates are Lake County: SW.1/4, SW.1/4, SE.1/4, Sec. 13, T.5N, R.1W. The site (fig.2) is a vacant lot near an industrial park in suburban north Hammond, and will soon be a construction site. Fossil remains were found in place in Paleocene terrace material, immediately below a shallow brown loam soil and overlying the Rolling Meadow Formation (Miocene sands).

In fact, all of these revisions are wholly subservient to the central imperative of the scientific style: the support of a privileged ontology. The imperative to enlist the possibilities of grammar in support of this privileged ontology also explains the scientific preference for the passive voice: it is a routine means for making physical objects and events the subjects of scientific sentences. No matter that the passive is an alternative voice that cannot be used consistently without clumsiness; no matter that the overuse of complex noun phrases interferes markedly with comprehension.[5]

Because it privileges ontology over felicity, scientific prose may be clear, yet hard to read; well-turned, yet barbarous. A typical paragraph from the series of cancer papers follows, one that relies on a syntax of complex noun phrases and the systematic overuse of the passive voice. The research it reports is part of a program that concerns the interruption of a biochemical process in the cell, activated by a Na^+K^+ pump. As a result of this interruption, enzymes called protein kinases form a chain reaction, a cascade. This reaction causes an anomaly common to the cancer cell, increased glycolysis, the overproduction of acid from carbohydrate:

> The results of two immunological experiments support the homology of EAT-mouse PK_F and avian sarcoma virus $pp60^{src}$. In the first we compared antisera to purified $EAT-PK_F$ and sera from rabbits bearing RSV-induced tumors (TBR sera) for their ability to precipitate viral $pp60^{src}$ from RSV-transformed cells. Rat cell lines transformed with three different strains of RSV (SR-NRK, B77–NRK and PrC-NRK), and as controls nontransformed NRK cells and NRK cells transformed by the unrelated Kristen sarcoma virus, were labeled with ^{35}S-methionine. The cell lysates were divided into aliquots, treated either with anti-PK_F serum or with TBR serum, and then the immune precipitates were collected by adsorption to formalin-fixed Staphylococcus aureus cells. The extracts after the removal of the Staphylococcus were then treated with the alternate sera, either TBR or anti-PK_F, respectively, and any immune complexes were collected. A fluorogram of the dissolved immune precipitates after electrophoresis on an SDS polyacrylamide gel is shown in Figure 4. (Spector, Pepinski, et al. 1981, pp. 11–12)

Transformed into the active voice, with their noun phases considerably reconstructed, the last three sentences might read: "We divided the cell lysates into aliqouts treated with one of two serums: anti-PK_F or TBR. At the same time, we collected the immune precipitates by adsorption to Staphylococcus aureus cells fixed with formalin. After we removed the Staphylococcus, we treated the extracts with either TBR or anti-PK_F, and collected any immune complexes. In Figure 4, we see a fluorogram of the dissolved immune precipitates after electrophoresis on an SDS polyacrylamide gel." In these revised sentences, the scientists ("we") and their manipulations (strong verbs in the active voice) exist on an equal standing with the physical objects they manipulate.

The Role of Tables and Figures

Tables and figures serve scientific argument in two ways:[6] first, by bringing the reader closer to the experience that grounds the argument, they add semantic weight to its terms; second, they suggest relationships, ideally causal relationships, among the physical objects whose behavior they summarize. Tables and figures can perform these functions only if the physical objects whose properties they display exhibit invariance.

Scientific illustrations, an important subset of figures, may consist of drawings, schematics, camera obscura tracings, or electron micrographs. Whatever their form, it is crucial that, through every alteration in their scale, every change in their angle of view, these illustrations maintain the identity of the physical objects they purport to represent. This is a triumph of projective geometry, of perspective: indeed, "there are few sciences . . . that are not predicated in one way or another upon this power of invariant pictorial symbolization" (Ivins 1938, p. 13).

Invariance extends beyond pictorial symbolization. In a table or a graph, a physical object may be variously represented: it may be a number, the height of a bar, a point on a curve. But these changes in representation must never alter the identity of the object displayed. This is a triumph of simplification, of normalization, and of the apotheosis of normalization, quantification: not mice, but their brain cells; not the unique features of particular cells, but what all cells have in common, a commonality realized perhaps in moles per minute per milligram, perhaps in ratios, perhaps in chi squares. In tables and figures most of the properties of the actual physical objects, of mice or of men, have been discarded, and all that remain have been normalized, ideally through quantification.

By means of normalization and quantification, tables and figures perform a crucial ontological role in the transformation of the terms of science into theoretically important physical objects and events. In this respect, the tables and figures in the series of cancer papers discussed earlier are typical. In Table 1, the labeling of rows and columns reinforces the ontological stability of the sources of Na^+K^+ ATPase, of the reagent, and of measures that operationally define the relative efficency of the Na^+K^+ pump, the focus of concern. At the same time, the numerical arrays accomplish two purposes: they add semantic weight to the changes that take place within this ontologically stable framework, and they support the hypothesis that pump efficiency is causally implicated in cancer.

These changes are also the focus of Figure 2, a plot of an expanded set of the data in the last column of Table 1, Experiment 1. In this graphic assertion of the relative efficiency of the pump, the quantitative information encrypted in each of nineteen data points is fully recoverable. But what stands out is the prominent curve, a symbol of the central, hypothesis-confirming relationship: the effect of quercetin on pump efficiency. Figure 2 has the further semantic purpose of all data graphics; it "links . . . two variables, encouraging and even imploring the viewer to assess the possible causal relationship between the plotted variables. It confronts causal theories that X causes Y with empirical evidence as to the actual relationship between [in this particular case, quercetin and pump efficiency]" (Tufte 1983, p. 47).

Figure 3, a photograph transformed into a data graphic by its superimposed grid, provides further visual support for a causal theory. The markers in lane 5, visually the most prominent, simultaneously depict and encode the claim that "the purified PK_L-phosphorylating activity from the B77–NRK cells also migrates as a single polypeptide of 60 kd [kilodaltons]" (Spector, Pepinski, Vogt, and Racker 1981, pp. 10–12). This claim favors the key hypothesis that the kinase cascade is causally implicated in cancer.

Figures 2 and 3 are members of a suite of graphics that culminates in Figure 4, a set of models for the operation of the kinase cascade. In this diagram we see the culmination of Racker's research program:

Table 1 Na$^+$/ATP ratios of vesicles reconstituted with Na$^+$K$^+$-ATPases from tumor, brain, and electric eel

In Experiment 1, reconstitution was performed with 80 µg of the ATPase preparation in the presence of 2.8 mg of asolectin in 50 mM imidazole-H$_2$SO$_4$ (pH 7.5), 75 mM K$_2$SO$_4$, 50 mM Na$_2$SO$_4$, 20 mM 2-mercaptoethanol, and 2% octylglucoside, in a final volume of 0.12 ml. After 15 s at 0°C, 3 ml of the same cold buffer (without octylglucoside) was added. Centrifugation at 48,000 rpm (144,000 × g) for 30 min in a Ti-50 rotor of a Spinco centrifuge yielded a pellet which was suspended in 0.15 ml of the same buffer, and 50 µl were used for each assay. Quercetin solutions (1 to 3%) were prepared in dimethylsulfoxide and kept at 0°C in the dark. Controls were run with the same amounts of dimethylsulfoxide. Experiment 2 was performed the same as Experiment 1 except that 1.4% deoxycholate was used instead of 2% octylglucoside.

Source of Na$^+$K$^+$-ATPase	Quercetin	^{22}Na uptake			ATPase	Na$^+$/ATP ratio
		-ATP	+ATP	Δ		
	µg/mg lipid + protein	*nmoles min^{-1} mg protein^{-1}*			*nmoles min^{-1} mg protein^{-1}*	
Experiment 1: octylglucoside dilution method						
Ascites tumor	0	471	985	514	1510	0.34
	6	455	965	510	1420	0.36
	10	407	955	548	822	0.67
	16	466	1017	551	642	0.86
	20	436	942	507	402	1.26
	24	530	924	394	275	1.43
	30	422	682	260	275	0.95
Electric eel	0	418	798	380	210	1.81
	24	426	701	275	156	1.76
Experiment 2: deoxycholate dilution method						
Ascites tumor	0	483	993	510	1509	0.34
	16	476	973	497	495	1.00
	24	491	796	305	229	1.33
Mouse brain	0	486	978	492	295	1.67
	16	463	899	436	248	1.76
	24	475	771	296	179	1.65

Source: From Spector, O'Neal, and Racker 1980b, p. 5506. Reproduced by permission of the American Society for Biochemistry and Molecular Biology.

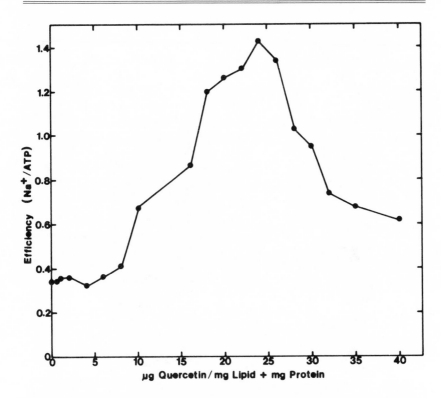

Figure 2 Effect of quercetin on the sodium-pumping efficiency of the reconstituted Na^+K^+-ATPase. The experimental conditions were as described in the legend to Table 1 for Experiment 1. (From Spector, O'Neal, and Racker 1980b, p. 5506. Reproduced by permission of the American Society for Biochemistry and Molecular Biology.)

a causal analysis of the aberrant cell activity and growth that is cancer. In each case "a gene product, either a kinase itself or an activator of a kinase, starts a cascade of phosphorylation which increases glycolysis and other metabolic pathways and alters components of the cytoskeleton which in turn affect structure and membrane processes such as ion transport" (Racker and Spector 1981, p. 306).

Quine calls attention to the "ironical but familiar fact that though the business of science is describable in unscientific language as the discovery of causes, the notion of cause itself has no firm place in science" (1966, p. 242). But one should not generalize from the low

A   **B**

Ehrlich   **B-77**

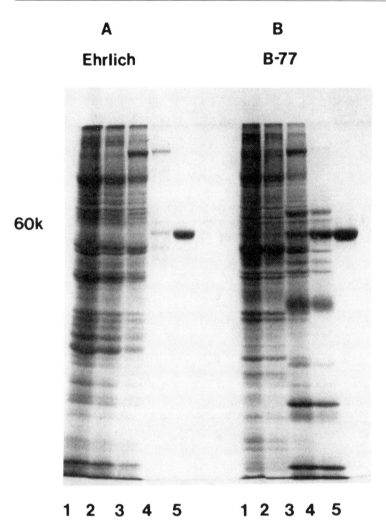

60k

1 2 3 4 5   1 2 3 4 5

Figure 3 Polypeptide profiles during purification of PK$_F$ and pp60^{src}. Aliquots of partially purified PK$_F$ from EAT cells and partially purified pp60^{src} from B77-NRK cells were subjected to electrophoresis on SDS-polyacrylamide gels. The gels were stained with Coomassie blue and photographed with a Wratten #99 filter. (A) purification of EAT PK$_F$. Lane 1, crude cytoskeletons (50 μg); Lane 2, NP-40 extract (35 μg); Lane 3, active fraction from octyl-Sepharose (35 μg); Lane 4, active fraction from casein-Sepharose (10 μg); Lane 5, PK$_F$ after isoelectric focusing (15 μg). (B) purification of B77-NRK pp60^{src}. Lane 1, crude membranes (50 μg); Lane 2, NP-40 extract (35 μg); Lane 3, active fraction from PK$_L$-Sepharose (35 μg); Lane 4, active fraction from casein-Sepharose (25 μg); Lane 5, pp60^{src} after isoelectric focusing (20 μg). (From Spector, Pepinsky, Vogt, and Racker 1981, p. 12. Reproduced by permission of Cell Press.)

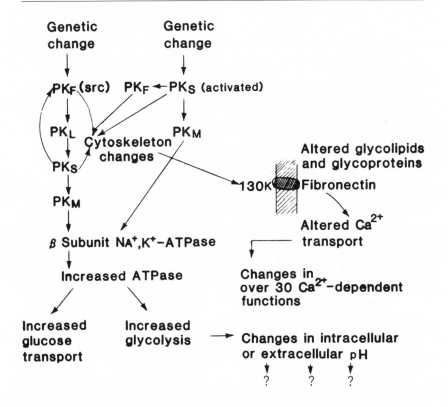

Figure 4 Models of transformation. The model on the right, altered Ca^{2+} transport, links some related research. (From Racker and Spector 1981, p. 306. Reproduced by permission of *Science*.)

visibility of causality in theoretical physics to its lack of importance in science as a whole. Scientific prose and its accompanying tables and figures work together toward a single metaphysical end: to create a world of quantified, causally arrayed physical objects. For working scientists, science continues to be the "accommodation of language to the causal structure of the world" (Boyd 1979, p. 358; emphasis deleted).

In science, then, tables and figures combine with text "to channel the judgment" (Bastide 1985, p. 147), to mobilize all the means of persuasion in the interest of a particular cause. By raising the price of dissent, tables and figures work together with text to win scientific arguments: behind the text, there are numbers and curves; behind numbers and curves are laboratory procedures and the equipment that makes these procedures possible. Bastide puts it well: "I believe exclusively neither in mathematization nor in linguistification! What matters is to be convincing; whatever is convincing is fine. But some ways are better than others: the economical use of graphic space that allows us to see the result at a glance, and the purely mechanical movement of scientific instruments (l'automatisme), which serves to guarantee "reality" (1985, p. 151; my translation).

The Problem of Metaphor

Although the management of scientific prose seems to support a metaphysical realism, it is impressions only that are being managed. If science really involved the accommodation of language to the causal structure of the world, scientific prose would name, not misname, physical objects. But science is full of metaphor, and it is the nature of metaphor deliberately to misname. Moreover, science cannot do without these "semantically bizarre sentences";[7] it is universally recognized that metaphor is indispensable to science.

Metaphysical realism can recognize the value of metaphor in scientific discovery, but it must deny metaphor a role in the knowledge a mature science professes. To this traditional view, Richard Boyd gives new strength and coherence. Although for Boyd scientific terms and metaphors begin alike as heuristics useful in scientific discovery, they differ not in the way that they profess to refer but in the categories of their professed referents: terms like "californium" point in the direction of natural kinds; metaphors like the genetic code, in the direction of natural relationships. Moreover, the terms and metaphors of science do not refer by definition, by specifying the necessary and sufficient conditions of their referents. Rather, each term refers by virtue of its constant association "with a sample or exemplary causal effect of its referent, or with a description of stereotypical samples or causal effects" (1979, p. 366). And in each case reference is provisional, a promissory note redeemable by future research that will determine what is the case.

Because natural relationships are central to research programs, metaphors are central to the doing of science: they "are (at least for a while) open-ended precisely because the research program they help to initiate is incomplete" (Boyd 1979, p. 370). But "if the metaphors are apt, and if they are drawn in sufficient detail, the differences in functional (relational) properties of the literal referents . . . will serve—by analogy—to disambiguate the referents of these terms in their theory-constitutive metaphorical applications" (p. 369). As theories mature, as their descriptions approach closer to the causal structure of the world, metaphors will disappear, to be replaced by wholly referring terms (p. 362).

But metaphors in science can disappear only when scientists can redescribe natural relationships in language free from metaphor. And they cannot; the cancer research papers typify this inability. Underlying their science is a way of talking that turns living cells and processes into their presumed mechanical and computer counterparts: a unit of the living cell is a "pump," and a biochemical process is a "cascade" that embodies a "positive feedback mechanism" (Racker and Spector 1981, p. 305). Racker's research program does not, cannot, aim at eliminating all reference to the machine and the computer; it is designed rather to specify exactly what sort of machine or computer a biological object really is. This sort of explication only serves to ensure the permanent entrenchment of metaphor. Initially, scientists imagine physical objects by thinking of them as machines or computers. Since they frame their causal hypotheses in these terms, it need come as no surpise that, in time, these objects seem more and more to *be* machines and computers.

The Problem of Reference

That the kinase cascade was a notorious scientific fraud may seem to undermine the argument of the previous sections, an argument that emphasizes on the typical nature of the papers of Racker's team. But the contrary is true: the reproduction of such typicality was essential to Spector's deceit and guarantees the authenticity of his rhetoric. The fraud is typical in another way important to this chapter: its exposure typifies the defeat of all ontological claims in science, whether fraudulent or not, a defeat that reveals the social and linguistic nature of such claims.

A necessary condition of doing science is "systematic ontological equivocation" (Hull 1988, p. 153): scientists treat their successful experiments and observations as demonstrations of a true theory of natural kinds. In this way, each scientist acts as a representative of the relevant epistemic community that automatically reads such experiments and observations as demonstrations of natural regularities. In Hull's words, science is an activity in which "term-tokens are tested and transmitted locally but interpreted globally as types" (1988, p. 153; see also p. 141 and Gross 1988). Experiments and observations not so interpreted lack not only generality but the requisite objectivity, an objectivity that can be bestowed only by the relevant epistemic community: "The sort of objectivity that gives science its peculiar character is the property of social groups, not isolated investigators" (1988, p. 127). In other words, to say that a scientific term successfully refers is to say no more than that social and linguistic methods bear out the existence of its purported referent; to say that a scientific term failed to refer is to say no more than that these same methods failed in this same task. In each case, social and linguistic practices are the criterion of ontological claims.

In the case of the kinase cascade, Racker's suspicions were aroused after a younger colleague "discovered that the data obtained from an experiment . . . were incompatible with the experimental protocol" (1981, p. 1313): autoradiograms, supposed to be coded with radioactive phosphorus, were marked instead with radioactive iodine. This discrepancy led to a withdrawal of papers and seminar claims. But it did not result in the abandonment of the kinase cascade research program—quite properly. The program was abandoned only after the persistent failure of replication: no one could successfully repeat most of the experiments on which the kinase cascade theory was based—not Spector, who was responsible for the suspected fraud, not Racker after two years of hard work.[8]

The causal structure of the world to which scientific terms refer is not a physical object; it is not a relation; it is not a process. Instead, it is a hypostatization of a set of social and linguistic practices, the source of which is a pair of experienced hands, a synecdoche for the technical and observational proficiency that, finally, divides scientific truth from error: "I have to go back to square one," [Racker said]. "I will not believe anything Mark [Spector] did until I repeat it with my own hands" (Kolata 1981, p. 316). The real Racker-Spector story is one of manipulation, in the root sense of that word: hands built this "castle in the clouds"; hands would tear it down. Spector had "golden hands," but when Racker tested Spector's enzymes, "in [his] hands, they were completely negative" (Kolata 1981, pp. 317–318).

Through the language they use, through the tables and figures they construct, scientists encourage the inference that the desideratum of their claims is the causal structure of the world. But to refer is merely to link the social and linguistic practices that constitute causal language with those that constitute experimental or observational work. To refer is to name a physical object but to mean an experimental or observational event; to impute causation is to name a causal relationship among physical objects but to mean a correlation between experimental or observational events.

In the following passage Racker's mastery of the stylistic subtleties inherent in natural language achieves, almost imperceptibly, the necessary translation: the manipulations that constitute his work become, through language, the causal relationships whose truth he wishes to assert:

> The presence of PK_M in the brain raised the question of how its activity is controlled. In fact, according to radioimmunoassays all four kinases of the cascade are present in the brain, yet the Na^+, K^+-ATPase does not become phosphorylated. It appears that the cascade is kept under control by a 6000–dalton polypeptide . . . that inhibits PK_S activity. Why should nature invent a protein cascade that impairs the efficiency of a pump and then keep it under control? An obvious answer is that the cascade has another purpose and the phosphorylation of the β subunit is a biological by-product. (Racker and Spector 1981, pp. 305–306)

The first two sentences of this passage concern laboratory manipulations; in the third, causation supervenes: the polypeptide is personified. In the final two sentences, nature herself boldly appears, the overseer and rational manipulator of the mechanism she invents.

To scientists, the tacit shift from social and linguistic practices to causal phenomena is routine—a condition of their employment, a condition, indeed, of doing good science. But for those of us who are outsiders, to leap from the social and linguistic practices of the laboratory to a universe of causally arrayed physical objects is simply to mislay Occam's razor.

Ideology is a generally notorious concept. We think of *Triumph of the Will:* the Wagnerian strains, the massed Nazi flags, the crowds surging under Hitler's gaze. But ideology need be neither malevolent nor antirational. When everyday existence becomes genuinely problematic, puzzlement is our natural state. We ask: what is real? what is a fact? what is a rule? In this uncertainty, an uncertainty that science routinely creates, ideology naturally flourishes, a map of problematic social reality, a matrix for the creation of collective conscience (Geertz 1973, p. 220). Although the style of science encourages us to infer that a family of disciplines has privileged access to the causal structure of the world, problems with metaphor and reference effectively block that inference. Style in science is not a window on reality, but the vehicle of an ideology that systematically misdescribes experimental and observational events. It is their ideological stance that makes contemporary scientists the legitimate heirs of medieval theologians: theirs is not a dispassionate search for truth, but a passionate conviction that the truth is their quotidian business.

In science, understandably, metaphor is this ideology's chief tool: at the scientific verge, words routinely fail to refer. In discovery, metaphor is prominent; in verification, it remains so: scientists have no better resource for expressing what they do not see, the universal mechanism whose audible and visible traces they purport to track.

CHAPTER 6

The Arrangement of the Scientific Paper

Reading experimental or descriptive papers in science, we invariably experience an inductive process, a series of laboratory or field events leading to a general statement about natural kinds; in theoretical papers we experience the opposite movement, a series of deductions whose conclusions invoke or imply confirming observations. In this chapter I point out that the principal motive behind these regularities in arrangement is epistemological: they reenact the scientists' faith in the existence of a suite of methods by which the causal structure of the world can be displayed, directly or indirectly, to the senses.

The Experimental Report as a Baconian Induction

The substance of the experimental report has been characterized as "an *a posteriori* rationalisation of the real process."[1] Its "residual description," it has been asserted, "is based on typification. The sequence of steps outlined in the report is a normalised, average depiction in which many of the peculiarities and exactitudes of the laboratory have been omitted or transformed."[2] The social scientists who so characterize the experimental report see its content as the result of a social process, the product of "a fierce fight to *construct* [social] reality" (Latour and Woolgar 1979, p. 242). But the form of the experimental report is even more obviously a typification—a philosophical typification, as P. D. Medawar (1964) notes. Medawar finds the source of this typification in the inductive theory of John Stuart Mill; I believe that it reaches farther back—to the writings of Francis Bacon, where we find the philosophical beginnings of modern experimental science: the arrangement of the experimental report is a realization of the principles of Baconian induction.

According to Medawar, this arrangement unreasonably distorts the process of experimental science; in accordance with the hypothetico-deductive method, to him the method of science, Discussion should precede Introduction rather than follow Results. But the order of these sections, however much it may distort the processes of science—however much it may fail to reflect contemporary notions of these processes—persists for good reason: the arrangement of the experimental report recapitulates a movement from the contingency of laboratory events to the necessity of natural processes; in other words, the arrangement of the experimental report reenacts the process of induction. This reenactment satisfies a recurrent need to justify the enterprise of experimental science in the face of the problematic nature of the inductions on which that science relies for the creation and certainty of its knowledge.

To illustrate, I will analyze two experimental records typical of best practice.[3] These were written almost exactly three hundred years apart. The first, in physics, was set down nearly at the inception of modern experimental science; the second, in biology, was reported only a quarter of a century ago. Robert Boyle's record of 1662 claims that at constant temperatures the volume of a gas is inversely proportional to the pressure exerted upon it, a phenomenon known generally as Boyle's law.[4] Marshall W. Nirenberg and J. Heinrich Matthaei, in a paper published in 1961, resolve a crucial problem in molecular biology: the nature of the genetic code. Despite their considerable differences in style, subject matter, and technical sophistication, these two reports may be profitably viewed as instantiations of Baconian inductive principles: records of a series of luciferous experiments, "Experiments of Light . . . by which the understanding may determine on the true causes of things" through "true induction" (1937, p. 372; 1960, pp. 151–152; see also p. 96). And this is true not just in the general case, as has been asserted (Kronick 1976, p. 279), but section by section.[5]

Introduction. According to Bacon, experimental science progresses "by varying or extending the same experiments [and] by transferring and compounding divers experiments the one into the other" (1962, p. 100). A research program, in fact, consists of "a double scale or ladder, ascendent and descendent; ascending from experiments to the invention of causes, and descending from causes to the invention of new experiments" (1962, pp. 90–91). Boyle's introduction and that of Nirenberg and Matthaei conform closely to the Baconian model: each report indicates that the experiments it will relate are part of a research program opening out routinely to further related experiments. Boyle states: "We shall now endeavour to manifest by experiments purposely made, that the spring of the air is capable of doing far more than is necessary for us to ascribe to it, to solve the phaenomena of the Torricellian experiment" (1965, p. 337). Nirenberg and Matthaei are rigorously programmatic: their Introduction locates their report within their own program of research and within the larger effort to break the genetic code (compare the discussion by Markus 1987, pp. 36–39).

Methods and Materials. The section on Methods and Materials also has clear Baconian roots. For Bacon, it was not enough that the experimenter was able to obtain a given result; all men should be able to so, that they "may be free to judge for themselves whether the information obtained . . . be trustworthy or fallacious" (1960, p. 282; see also 1937, p. 372). Boyle pleads for "plain and easy experiments" because they are "most easy to be tried . . . and to be judged of, both in relation to their causes and their effects" (1965, p. 277; see also pp. 287 and 343; see also Shapin 1984). In contemporary science, replication is no longer routine. Yet Nirenberg and Matthaei routinely detail their methods just as Boyle did. In contemporary reports the possibility has replaced the fact of replication, but the epistemological grounding is the same: only those experiments that are replicable truly illustrate natural laws.

Results. In Boyle and in Nirenberg and Matthaei, the section on Results follows Methods and Materials; in both sections, as in Baconian induction, experimental results are revealed as the foundation of scientific knowledge: "The passages and variations of nature cannot appear so fully in the liberty of nature, as in the trials and vexations of art" (Bacon 1962, p. 73; see also 1964, p. 99). Therefore, we must put nature "on the rack";[6] we must engage in experiments guided by the two key principles of induction: exclusion and concomitant variation. Exclusion eliminates possible, but mistaken, lawlike relationships; it is "the rejection or exclusion of the several natures which are not found in some instance where a given nature is present, or are found in some instance where a given nature is absent, or are found to increase in some instance when the given nature decreases or to decrease when a given nature increases" (1960, pp. 151–152).

The second principle, the obverse of exclusion, is concomitant variation: "to find such a nature as is always present or absent with the given nature, and always increases and decreases with it, and which is . . . a particular case of a more general nature" (1960, pp. 151–152).

In both Results sections, these Baconian principles are manifest. In Nirenberg and Matthaei, exclusion is partly a product of the experimental design, partly a rejection within the experiment of initially plausible, finally unsuccessful chemical affinities. Concomitant variation produces for these two molecular biologists the following arresting comparison: in stimulating the incorporation of L-phenylalanine into a protein resembling poly-L-phenylalanine, polyuridylic acid is *almost five hundred times* more successful than its nearest rival. In the case of Boyle, the carefully described experimental apparatus itself excludes. The concomitant variation is inverse, a lawlike relationship seemingly unanticipated by Bacon. This relationship stands out with startling clarity in Boyle's table (adapted from Boyle 1965, p. 340):

Pressure	Actual volume	Anticipated volume
5	$70\frac{11}{16}$	70
10	$35\frac{5}{16}$	35
3	$117\frac{9}{16}$	$116\frac{4}{8}$
6	$58\frac{13}{16}$	$58\frac{2}{8}$
12	$29\frac{2}{16}$	$29\frac{2}{16}$

Discussion. Bacon and his followers represented the inductive method as a conscious break with scholastic science, whose allegedly meager success they attributed to a radical imbalance: a distrust of sensation coupled with an overreliance on speculation. Bacon held that scholastic scientists regularly deduced new scientific conclusions from premises insufficiently founded, as all the premises of natural science must be founded, on the evidence of the senses. On inductions from such sensory evidence, says Bacon, "are centred all our hopes. This is the method which by slow and fruitful toil gathers information from things and brings it to the understanding" (1964, p. 89). If, in the enterprise of experimental science, sensation and speculation could be equal partners, "if the two could be joined in a closer and holier union the prospects of a numerous and happy issue are bright indeed" (1964, p. 98; see also 1960, p. 93).

These prospects were especially bright, in Bacon's view, because "in nature . . . principles are examinable by induction . . . and besides, those principles or first positions have no discordance with that reason which draweth down and deduceth the inferior positions" (1962, p. 211; see also 1964, p. 82). As a result of this ultimate concord, "a solid theory may in the process of time be superstructed" by induction, a theory from which deductive conclusions may be reliably drawn.[7] By induction "generalisations lying close to the facts may first be made, then generalisations of the middle sort, and progress thus achieved up the successive rungs of a genuine ladder of the intellect" (1964, p. 99).

This curbing of speculation, this emphasis on "generalizations lying close to the facts," is equally apparent in the Discussions of Boyle and of Nirenberg and Matthaei. Boyle never actually enunciates Boyle's law. Concerned with discrepancies between actual and expected results, discrepancies he attributes "to some such want of exactness as in such nice experiments is scarce avoidable (1965, p. 341)," he will say only that "common air, when reduced to half its wonted extent, obtained near about twice as forcible a spring as it had before: so this thus compressed air being further thrust into half this narrow room, obtained thereby a spring about as strong again as it had, and consequently four times as strong as that of the common air" (pp. 341–342). Nirenberg and Matthaei's conclusion also clings close to laboratory observations: "The results indicate that polyuridylic acid contains the information for the synthesis of a protein *having many of the characteristics of* poly-L-phenylalanine." Their speculations are clearly curbed by their modality: "One or more uridylic acid residues therefore *appear to be* the code for phenylalanine. Whether the code is of the singlet, triplet, etc., type has not yet been determined" (1961, p. 1601; my emphasis).

Sequence of the Sections. The sequence of the sections of the experimental paper also has Baconian roots: a steady march from Introduction to Discussion, from the contingency of laboratory events to the necessity of natural processes.[8] This order is, as Woolgar aptly states, "a picture of the discovery process as a path-like sequence of logical steps toward the revelation of a hitherto unknown phenomenon" (1981, p. 263). Experimental reports begin by placing the incidents they report in the context of a research program whose goal is the discovery of natural laws:[9] introductions recreate a theoretical world in which the otherwise contingent events of the laboratory will attain their significance as scientific experiments, instantiations of particular natural laws. In Boyle's case, the local manipulations of an air pump will become examples of a universal relationship between the pressure and volume of gases (1965, p. 337); in Nirenberg and Matthaei, the laboratory manipulations of a "stable, cell-free system" will become examples of the operation of the genetic code, the ground-plan of all living organisms (1961, p. 1588).

The case against contingency is supported by two apparent violations of the norms of communication. Experimental reports seem to violate the maxim of quantity ("Do not make your contribution more informative than is required"): Methods and Materials and Results each provide more detail than is necessary merely to follow the course of an experiment.[10] But in fact the maxim of quantity is strictly adhered to: the purpose of these two sections is not to enable understanding but to permit replication, the confirmation that nature, not the experimenter, is the cause of laboratory events.

Apparently the maxim of relation is also violated ("Be relevant"):[11] both in Boyle and in Nirenberg and Matthaei, events have a "so what" quality that would be fatal to ordinary communicative purposes. In Nirenberg and Matthaei, this involves the behavior of certain organic molecules; in Boyle, that of quicksilver: "we ob-served . . . that the quicksilver in that longer part of the tube was 29 inches higher than the other" (1965, p. 337). But the communicative purposes of these reports are not ordinary: necessary, not contingent, events are being related. In the Discussion the significance of these events, their relevance, will be understood by framing them in the theoretical perspective of the Introduction. In other words, in the Discussion the data from Results will be transformed into candidate knowledge by adducing in their favor their close correspondence to the report's claim (Bazerman 1981, p. 366).

We should notice, as experimental reports routinely do not, that data and claims are from different worlds: data, from the laboratory; claims, from nature. In Methods and Materials and Results, replication, though it did not depend on any particular human being, depended nevertheless on some human intervention. In the *Discussion*, replication is seen ultimately to depend not on the scientist but on the lawfulness of nature: the laboratory experience has become a synecdoche, a natural index.[12] At this point, if the authors of reports have succeeded, their assertions have attained the status of facts and can be separated from the laboratory events from which they arose. It is this separation that makes the Abstract possible: at the conclusion of experimentation, laboratory events can be discarded in favor of their eternal sentences, the summarized relationship between natural kinds and natural processes.[13]

The Idea of Induction

Throughout its history, then, the experimental report has remained stubbornly ritualized, essentially unchanging, a realization in detail of the principles of Baconian induction, the order of its sections dramatizing a steady march from contingency to natural necessity. Developments in the philosophy of science, as P. D. Medawar notes, have been resolutely ignored. Why? Because this detailed realization in its particular sequence transforms experimentation into a myth about induction: the myth that inductive science is philosophically unproblematic, that it can lead directly from sensory experience to reliable knowledge about the natural world.

That induction with these capabilities is a myth has been known at least since the late Middle Ages. Grosseteste was not alone when he asserted that natural science offered its explanations "probably rather than scientifically." Nor was skepticism confined to the late medieval period. In the sixteenth century Nifo declared: "The science of nature is not a science *simpliciter* . . . that something is a cause can never be so certain as that an effect exists . . . for the existence of an effect is known to the senses. That it is a cause remains conjectural" (Crombie 1959, II, p. 16; see also pp. 32–34; 26–27). No serious defense of induction can avoid these arguments, yet Bacon avoided them. Boyle, a follower of Bacon, was ambivalent. His equivocation concerning the highest level of scientific hypothesis typifies his attitude: he asserts "that it be the *only* Hypothesis that can Explicate the Phaenonoma, or *at least*, dos explicate them so well" (1965, p. 135; my emphasis).

The problem of induction is really two problems. Insofar as the purported cause cannot be established independently of its effects, inductive reasoning is circular; insofar as induction moves from effect to purported cause, it commits the fallacy of affirming the consequent. Bertrand Russell says it well: "Domestic animals expect food when they see the person who usually feeds them. We know that all these rather crude expectations of uniformity are liable to be misleading. The man who has fed the chicken every day throughout its life at last wrings its neck instead" (1974, p. 21). The chicken argues validly: if the man cared for him, he would feed him; however, the chicken commits the fallacy of affirming the consequent when he argues: the man feeds him; therefore, he cares for him. In addition, when the chicken asserts that all acts of feeding are acts of caring because each one is, he is arguing in a circle. But the chicken is oblivious, and for good reason: as the days pass, evidence in his favor seems to accumulate steadily; the abundance of confirming instances regularly increases the probability that he is right—increases it, that is, until his last day.

Though it remains problematic, induction has never lacked serious defenders. Descartes, an early champion, tries to break out of the circle. Concerning his conclusions in the *Dioptrics* and the *Meteors*, he says:

> It is my view as to the connection of my conclusions that, just as the last are proved by the first, which are their causes, so the first may in turn be proved from the last, which are their effects. It must not be thought that here I am committing the fallacy called by logicians a vicious circle; for the effects are for the most part known with certainty by experience, so that the causes from which I have deduced them serve not to prove but to explain them—must indeed, be themselves proved by means of them. (1954, pp. 55–56)

But even though we may know an experience with certainty, we cannot know with the same certainty, but only through the mediation of some purported cause, that that experience is an effect. Cohen and Nagel, with a caginess born of three hundred years of disputation, try not to escape but to embrace the circle; to them, there is "a difference between a circle consisting of a small number of propositions from which we can escape by denying them all or setting up their contradictories, and the circle of theoretical science and human observation, which is so wide that we cannot set up any alternative to it" (1974, p. 379).

My concern is not with resolving this quarrel; I want merely to note its presence: two parallel philosophical positions, each as old at least as the origins of modern experimental science, the one supporting, the other undermining its empirical claims.

The Theoretical Paper:
Einstein's Papers on Relativity

In the foundational achievement of modern theoretical science, the *Principia*, Newton asserts the deductive essence of his method, its inexorable drive from physical theory to sense experience: "From these forces [of gravity], by other propositions which are also mathematical, I deduce the motions of the planets, the comets, the moon and the sea" (1974, I, xviii). In the Euclidean arrangement of the *Principia*, this deductive essence is given its formal realization.

In his work and thought, Einstein exemplifies this Newtonian inheritance. In a typical paper, "Does the Inertia of a Body Depend upon Its Energy Content?" his axioms are (1) "the Maxwell-Hertz equations for empty space," (2) "the Maxwellian expression for the electromagnetic energy of space," and (3) the principle of special relativity. By means of Euclidian linguistic forms ("Let there be a stationary body in the system $\{x, y, z\}$ . . .), Einstein creates an ideal physical system. Its equations finally lead to the conclusion that "the mass of a body is a measure of its energy-content": $m = L/c^2$ or, in its more familiar guise, $E = mc^2$ (1952, pp. 67–71). Einstein derives this result by applying mathematics to an ideal physical system, but the conclusion itself is a law of nature displaying a new ontological equivalence: "The special theory of relativity has led to the conclusion that inert mass is nothing more or less than energy" (1952, p. 148).

In the penultimate sentence of this paper, Einstein moves without transition from the realm of physical theory to the experiential world, the only source for the confirmation of the new natural law he has derived: "It is not impossible that with bodies whose energy-content is variable to a high degree (e. g. with radium salts) the theory may be successfully put to the test."

Thus, "Does the Inertia of a Body Depend on Its Energy-Content?" is typical of Einstein's papers: a Euclidian deduction leads to a startling prediction of a hitherto unobserved effect of the physical laws that that deduction entails.[14] But what is predicted is not simply an effect of the equivalence of mass and energy; it is a revelation of the eternal character of nature. The effect is there whether or not we can measure it; it has always been there, whether or not there was anybody *to* measure it. The effect is a manifestation to the senses of the causal structure of the world; given the truth of the theory, mass and energy must be related in exactly the way the equations describe.

This revelation is an expression directly in accord with Einstein's philosophy of science, a philosophy in which physics is the search for laws that describe relationships among fundamental entities that completely determine physical events: "I still believe in the possibility of a model of reality—that is to say, of a theory which represents things themselves."[15] Although in no human enterprise is absolute certainty possible, near certainty certainly is: the laws of thermodynamics, for example, are unlikely ever to be overthrown.[16] Their near certainty established, determinate laws will unite into "a theory which describes the real state of things [*Dinge*] by a continuous function for which differential equations are valid" (Einstein 1959, pp. 86–87).

The verification of laws so fundamental is admittedly difficult. As physics develops, "the basic concepts and axioms distance themselves from what is directly observable, so that the confrontation of the implications of theory by the facts becomes constantly more difficult and more drawn out" (Einstein 1959, I, 27; see also 1954, p. 222). Moreover, "there is no logical bridge between the phenomena and their theoretical principles" (1954, I, 226). Nevertheless, scientific theories are possible: their existence depends only on the plausible conviction that the causal structures of the world are intelligible and rationally ordered: "Certain it is that a conviction, akin to religious feeling, of the rationality or intelligibility of the world lies behind all scientific work of a high order."[17]

The Idea of Deduction

The deductions of mathematical physics and the inductions of experimental science are alike in their presupposition of a naturally necessary link between theoretical formulations and sense experience, a link no logic can bridge. But logic is not the only difficulty. The bending of starlight in the vicinity of the sun, one of the classic verifications of general relativity, illustrates another: the problems of measurement and derivation that surround the verification of theoretical deductions. The measurement of stellar deviation is possible only during an eclipse; it is "the absolute measurement of a very small quantity during a particular short time interval under the usually quite difficult conditions of a temporary field-station in some more or less remote part of the world." Understandably, actual measurements have varied widely. Einstein had predicted a deviation of 1.7 seconds of arc; the actual results range from 0.93 to 2.73, a range so wide that the difference between extremes is larger than the predicted quantity.[18]

But even had these measurements converged, verification would not cease to be problematic. There is no question of applying general relativity directly, of measuring stellar positions, and plugging measurements directly into formulas. For the purpose of "calculational convenience" a set of idealized initial conditions is postulated at every stage; at each stage, when there is a failure of fit, the blame is placed not on the theory but "on the idealized initial conditions":[19] deviations count not against the theory, but against its verifiers.

After tabulating the results of various expeditions, Sciama says: "It is hard to assess their significance, since other astronomers have derived different results from a re-discussion of the same material. Moreover, one might suspect that if observers did not know what value they were 'supposed' to obtain, their published results might vary over a greater range than they actually do; there are several cases in astronomy where knowing the 'right' answer has led to observed results later shown to be beyond the power of the apparatus to detect" (Sciama 1959, p. 70; see also Collins 1975). This chain of inference suggests that the confirmation of scientific theories of great generality, theories far removed from the reality they purport to describe, is little endangered by experience—that the confirmation of such theories is as much a persuasive gesture as a decisive test.

When I speak of the scientific paper as a myth, it is Lévi-Strauss's use of the term that I have in mind. To Lévi-Strauss a myth is not merely a story; indeed, it is essentially not a story at all, but a "logical model capable of overcoming a [fundamental] contradiction" in life and thought.[20] Myths are designed to cope with contradictions so deep that they cannot be overcome. Rightly read, a myth is a record of temporary success, ultimate failure. In my view the scientific paper instantiates a myth designed to cope with the fundamental contradictions at the metaphysical heart of science. Each scientific paper exhibits terminological stability, the *sine qua non* of certain knowledge. Each seeks to persuade us that if we follow proper procedures we will establish reliable knowledge of the natural world. Each assures us that through theory we create an indissoluble link between sense experience and the transsensual world that is its cause.[21]

But the matter is not so simple: the set of all scientific papers undermines the myth that each paper instantiates. First, this set exhibits terminological instability, the *sine qua non* of opinion; second, it undermines the certainty of scientific knowledge, for what science routinely creates, it routinely overturns; finally, this set proclaims successive—and often contradictory—truths, a history hard to reconcile with any coherent notion of progress: light is a particle, light is a wave, light travels through ether, there is no ether.

Finally, we must differentiate between the philosophy of science of philosophers, and that of scientists. The latter had as its model the English experimental school of Boyle and Newton, a model inspired by Bacon (Crombie 1959, II, 319–320). This philosophy was designed less to explain the possibility of science than to justify its practice to its practitioners. It is precisely this philosophy that the arrangement of the scientific paper embodies: by means of its form alone, each scientific paper, each act of significant science, routinely conveys, along with its particular conclusions, the justification of the enterprise whose *raison d'être* is reaching such conclusions.

CHAPTER 7

Copernicus and Revolutionary
Model Building

Was the Copernican revolution essentially irrational, essentially not a matter of evidence and argument? Certainly, at its beginnings, the Copernican system was no better than the Ptolemaic as a tool for practical astronomy. Still, these early deficiencies, well known to modern historians, hardly justify Paul Feyerabend's view in *Against Method* that allegiance to the Copernican system in Galileo's day amounted to "a blind faith," supported not by evidence and arguments but *"by irrational means* such as propaganda, emotion, *ad hoc* hypotheses, and appeal to prejudices of all kinds"; by "Galileo's trickery." Yet Feyerabend's views on the Copernican revolution cannot be lightly dismissed. He has a firm grasp of its essential condition: "an entirely new world view containing a new view of man and of his capacities for knowing." He is also acutely aware that "developing a good theory is a complex process that has to start modestly and that takes time" (1975, pp. 153–154, his emphasis; p. 84; p. 152, emphasis omitted; p. 98).

Without endorsing Feyerabend's radical notion that the development of science is and ought to be essentially anarchic and fundamentally irrational, without endorsing Feyerabend's own rhetoric, it is possible fully to endorse—and, indeed, usefully to extend—one of his deepest and truest insights: that early in the development of a new scientific world view, at a time when a new theory is seriously underdetermined by the evidence, "style, elegance of expression, simplicity of presentation, tension of plot and narrative, and seductiveness of content become important features of our knowledge" (1975, p. 157).[1] Applied not to Galileo's *Dialogue* of 1632 but to the first published work on Copernicanism, Rheticus' *Narratio Prima*, published in 1540, three years before *De Revolutionibus*, Feyerabend's insight can show us how and why rhetoric operates at the inception of at least one scientific revolution.[2]

In this chapter I shall show how *Narratio Prima* does more than summarize heliocentric intricacies; in this work, Rheticus justifies his choice of the Copernican over the Ptolemaic hypothesis. To do so, he goes beyond argument and evidence. Naturally, he gives reasons that his training leads him to believe are scientific. But since, as a result of this very training, he cannot regard these reasons as sufficient for full conviction, he sets them within the frame of a conversion narrative: he unites argument and narrative into a single structure, a model for radical intellectual change.

This model, which I call rational conversion, was an ingredient essential to advances in astronomy in the late sixteenth and early seventeenth centuries. The new Copernicans shared with their master an austere and intransigent realism: the new astronomy must be a mathematically parsimonious system in exact agreement with accurate observations and in precise conformity with a correct physics. But the realization of this explanatory ideal was not possible until the advent of classical dynamics more than a century after Copernicus' death. Thus, in a period of transition, the model of rational conversion made the shift to Copernicanism psychologically tolerable: it facilitated change.

At the same time, as the work of Brahe, Maestlin, and Kepler makes clear, the model of rational conversion facilitated an apparently contrary purpose: the undermining of its own justificatory role. Because the narrative element in rational conversion was by definition unscientific, it became an irritant, a psychic itch that provided the continued impetus to change. The result was the eventual abandonment of all reliance on psychological models undergirding the espousal of the new paradigm. But it was not until Newton that the task was completed to the satisfaction of a sizable segment of the scientific world; only then did astronomy rest, in their view, on a fully scientific base. On this reading, *Narratio Prima* becomes the opening move in an eventually successful campaign to create the first modern exact science.

The Scientific Case for Heliocentricity

The radical nature of Copernicus' conclusion should not blind us to the conservative origin of his methods. His science is an "argument field," many of whose components are shared with his Ptolemaic fellows and predecessors. To describe these components "is largely a matter of describing the things its actors take for granted, their self-evident truths."[3] Copernicus shares with his scientific past three sorts of self-evidence: an astronomical theory must reason by means of Euclidean geometry; it must agree with the most accurate astronomical observations; and it must conform to a correct—essentially an Aristotelian—physics.

In *Narratio Prima,* therefore, Euclidian methods are central: "My teacher . . . has embraced the whole of astronomy, stating and proving [*docendo et demonstrando*] individual propositions mathematically and by the geometrical method" (Rosen 1959, pp. 109–110; Hugonnard-Roche 1982, p. 42).[4] But geometric reasoning is not sufficient: there must be exact agreement between theoretical conclusions and accurate observations. Concerning a discovery of his teacher, Rheticus says that it "is indeed worthy of the highest admiration, since it is achieved with such great and remarkable agreement" with observations (Rosen 1959, p. 121).[5] By means of geometry, then, Copernicus intends to harmonize in one system all of the observations available to him—both those he has made himself and those he has inherited from such great predecessors as Ptolemy (Rosen 1959, pp. 131–132).

But geometry and accurate observation are in themselves insufficient. Rheticus cites with approval Aristotle's distinction between the mathematician and the physicist, between a person concerned with abstract dimensions and quantities and one concerned with these only insofar as they are the properties of real bodies (Rosen 1959, p. 140; Aristotle 1929, pp. 116 ff.). Because the astronomer deals with the properties of real bodies, "the mathematician who studies the motions of the stars is surely like a blind man who, with only a staff to guide him, must make a great, endless, hazardous journey that winds through innumerable desolate places . . . God will permit him to try his strength for a period of years, that he may in the end learn that he cannot be rescued from threatening danger by his staff" (Rosen 1959, pp. 163–164). Thus it is only through its union with a correct physics that mathematics, coupled with observation, realizes its full power as a tool of physical astronomy, its ability to reflect in its own simplicity the actual simplicity of the universe which it describes: "Lest any of the motions attributed to the earth should seem to be supported by insufficient evidence, our wise Maker expressly provided that they should all be observed equally perceptibly in the apparent motions of all the planets; with so few motions was it feasible to satisfy most of the necessary phenomena of nature" (Rosen 1959, p. 161; see also pp. 137 and 149).[6]

It is Rheticus' point that the assumption of heliocentricity permits the parsimonious demonstration of astronomical truth: "If anyone desires to look either to the principal end of astronomy and the order and harmony of the system of spheres or to ease and elegance and a complete explanation of the causes of the phenomena, by the assumption of no other hypothesis [than heliocentrism] will he demonstrate the apparent motions of the remaining planets more neatly and correctly" (Rosen 1959, pp. 164–165).[7] In fact, "the assumption of the motion of the earth on an eccentric [leads to] a sure theory of celestial phenomena, in which no change should be made without at the same time re-establishing the entire system, as would be fitting, once more on proper ground" (Rosen 1959, p. 140).[8] But because the heliocentric hypothesis yields astronomical truth so readily, it is also the primary candidate for the central physical truth about the universe, a truth so overwhelming that "it seems impossible that the earth should occupy the center" (Rosen 1959, p. 137; Hugonnard-Roche 1982, p. 55).

Copernican astronomy has now met the goal of science as set out in the *Posterior Analytics:* "The hypotheses of my teacher agree so well with the phenomena that they can be mutually interchanged, like a good definition and the thing defined" (Rosen 1959, p. 186; Ross 1971, pp. 49–54; Aristotle 1975, pp. 64–65). With mathematics and physics as his guides, Copernicus has taken Plato's admonition seriously: he has ensured a "clear waking vision" for astronomy by giving an "account" of the "assumptions [hypothesesi]" of his system (Rosen 1959, p. 142; Hugonnard-Roche 1982, p. 58). We now have a total system "in agreement with the truth" (Rosen 1959, p. 193).

The Rhetorical Case for Heliocentricity

Brahe's *Astronomae Instauratae Progymnastmata*, Schöner's *Opusculum Astrologicum*, Clavius' *In Sphaeram Joannis de Sacro Bosco Commentarius*, *Astrolabium*, and *Gnomonices*, Fracastoro's *Homocentrica, sive De Stellis*, the anonymous summary of Peurbach's *Novae Theoreticae Planetarum* —the form of these typical Renaissance astronomical treatises is essentially that of Ptolemy's *Almagest*, itself clearly derived from

Euclid. Copernicus' *Commentariolus* also has a transparently Euclidean form, and *De Revolutionibus* is plainly imitative of the *Almagest*. But the overarching structure of *Narratio Prima* is storylike: the scientific case for Copernicanism is embedded in a framework of conversion.

Initially Rheticus presents Copernicus, his protagonist, as a great traditional astronomer, steeped in Ptolemy. Accordingly, in the first third of *Narratio Prima* the motion of the earth goes unmentioned, and Rheticus dwells upon those aspects of Copernicus' astronomy that do not require a discussion of heliocentricity: precession, the length of the tropical and sidereal years, the obliquity of the ecliptic, and lunar theory. In these pages Copernicus is presented as the heir to Ptolemy, his work written "in imitation of" the earlier astronomer (Rosen 1959, p. 131). As Ptolemy's heir, Copernicus relies on the most accurate observations—those he had made himself and those of his predecessors, especially Ptolemy—as a basis for an improved theory. In the first third of *Narratio Prima,* indeed, Copernicus is Ptolemy's partner in the restoration (*emendationem*) of their mutual science (Hugonnard-Roche 1982, p. 42). Ptolemy, the great predecessor, is mentioned more than fifty times, often with great praise: "Ptolemy's tireless diligence in calculating, his almost superhuman accuracy in observing, his truly divine procedure in examining and investigating all the motions and appearances, and finally his completely consistent method of statement and proof [*docendi et demonstrandi*] cannot be sufficiently admired and praised by anyone to whom Urania is gracious" (Rosen 1959, p. 131; Hugonnard-Roche 1982, p. 52).

But Rheticus' respect for his teacher exceeds his respect for Ptolemy, for Copernicus has "a burden greater than Ptolemy's," the burden of incorporating intervening astronomical observations into his system. Repeatedly referred to as *D[ominus] Praeceptor, D[ominus] Doctor Praeceptor,* or *D[ominus] Doctor*—Ptolemy is referred to only by his bare name—Copernicus creates a monument to observational astronomy, astronomical tables of really lasting value (Rosen 1959, pp. 131, 126).

In the midst of this portrayal of Copernicanism as the natural outgrowth of Ptolemaic astronomy, the crucial Ptolemaic assumption of geocentricity is suddenly and completely discarded. Near the end of a section on lunar theory, Rheticus says: "These phenomena, besides being ascribed to the planets, can be explained, as my teacher shows, by a regular motion of the spherical earth; that is, by having the sun occupy the center of the universe, while the earth revolves instead of the sun on the eccentric" (Rosen 1959, p. 135; Hugonnard-Roche 1982, p. 54). Since it comes without warning in a text up to then orthodox, this sentence was certainly a surprise to its sixteenth-century readers. It was also a shock: to believe it was to turn one's back on the central tenet of years of training, to discard completely the unquestioned assumption of respected colleagues and admired teachers, to desert forever the cosmos of fourteen centuries of astronomy. Little wonder that, for over a century, Copernicanism had relatively few significant adherents.

Sudden as this shift is, it is not without narrative preparation and does not long remain without argumentative justification: the conflict between Copernican and Ptolemaic astronomy is presented as a conflict within the scientific tradition and within Copernicus himself. While "walking in the footsteps of Ptolemy," Rheticus says, Copernicus nevertheless "became aware that the phenomena, which control the astronomer, combined with mathematics to compel him to make certain assumptions even against his wishes." He "had to assume" his hypotheses; he had to reach his conclusions "in obedience to the command given by the observations" (Rosen 1959, p. 186—translation emended; p. 140; p. 151; Hugonnard-Roche 1982, pp. 81, 57, 63). But, however radical, these were conclusions with which Aristotle would certainly have agreed: "I am convinced that Aristotle, who wrote careful discussions of the heavy and the light, circular motion, and the motion and rest of the earth, if he could hear the reasons for the new hypotheses, would doubtless honestly acknowledge what he proved in these discussions, and what he assumed as unproved principle." (Rosen 1959, p. 142).[9]

In an authentic search for truth, it is these unproved principles which, according to Aristotle's teacher, Plato, must be examined (Rosen 1959, p. 142; Hugonnard-Roche 1982, p. 58). Thus to deny the license to question geocentricity is to deny a central principle of the intellectual tradition in which Ptolemy worked, a denial that makes impossible any honest search for the physical reality behind celestial appearances.

In telling Copernicus' story, Rheticus is as sensitive to the value of contrast as is his great predecessor in conversion narrative, the author of *Acts*. The drama of rational conversion, which takes Copernicus from essentially the same observations as Ptolemy's to a diametrically opposed central conclusion, is intensified by the master's innate conservatism. The espousal of heliocentrism is especially reluctant, since "he is far from thinking that he should rashly depart, in a lust for novelty, from the sound opinions of the ancient philosophers, except for good reasons and when the facts themselves coerce him."[10] Even after he himself was entirely convinced of heliocentricity, we learn, Copernicus was reluctant to publish his results lest others be upset. Only the repeated urging of the Bishop of Kulm that there was no place for secrecy in science finally persuaded the great astronomer to publish not merely his tables,[11] but his whole system with proofs (Rosen 1959, pp. 187, 192–193).

In recounting Copernicus' conversion to heliocentrism, Rheticus is also recounting his own: "In my teacher's revival of astronomy I see . . . with both eyes and as though a fog had lifted and the sky were now clear" (Rosen 1959, p. 168; Westman 1975b). Throughout *Narratio Prima* Rheticus' own faith in his teacher's hypothesis increases slowly, a pace mirrored in his gradually increasing assertiveness. At first a notion by which the phenomena merely "can be explained," heliocentricity is soon afterward entertained more seriously: "Should we not attribute to God, the creator of nature, that skill which we observe in the common makers of clocks?" But the idea is still treated with a certain tentativeness: the argument is still full of words like "can," "suppose," and "as it would if" (Rosen 1959, pp. 135, 137; see also pp. 145, 150). No such hesitancy, however, attaches to this pronouncement near the end of the work: "That this covenant of earth and planet might be everlasting, God ordained that the first small circle of libration . . . should revolve once in the time in which one return of Venus to either of the movable nodes occurs" (Rosen 1959, p. 184; see also p. 185).

This increase in Rheticus' conviction is also mirrored in the gradual elevation of Copernicus' symbolic status: from heir to Ptolemy, to king, to general, to philosopher, to mythical hero: like Atlas, shouldering the world, or like Orpheus, rescuing the muse of astronomy from the underworld (Rosen 1959, pp. 131, 132, 150, 162–163, 164).[12] Such increasing mythical status is entirely appropriate, for rational conversion has led only to the discovery of the Neoplatonic truth: "self-caused," the sun is "the source of all motion and light"; "God stationed in the center of the stage His governor of nature, king of the entire universe, conspicuous by its divine splendor" (Rosen 1959, pp. 139, 146, 143; Kuhn 1981, pp. 128 ff.).

At this point, if one rereads or remembers anew the first portion of *Narratio Prima*, it will be apparent that crucial words—terms such as *hypothesis, theory, cause,* and *truth*—have changed their meaning. For example, a few pages before Rheticus' abrupt revelation of Copernicus' heliocentricism, this passage occurs:

> Nevertheless, the observations of all scholars and heaven itself and mathematical reasoning convince us that Ptolemy's hypotheses and those commonly accepted do not suffice to establish the perpetual and consistent connection and harmony of celestial phenomena and to formulate that harmony in tables and rules. It was therefore necessary for my teacher to devise new hypotheses,[13] by the assumption of which he might geometrically and arithmetically deduce with sound logic systems of motion like those which the ancients and Ptolemy . . . once perceived . . . and which careful observations reveal as existing in the heavens to those today who study the remains of the ancients. (Rosen 1959, p. 132)

In its original context, before the heliocentric revelation, this statement about new hypotheses can refer only to the technical innovations presented in the first third of *Narratio Prima*. The heliocentric revelation having occurred, the passage concerning the necessity of new hypotheses must be read rather as an announcement of a different order in the heavens, an order dictated by necessity. Of his teacher, envisioned in the first third of *Narratio Prima* as the ruler of the kingdom of astronomy, Rheticus had said: "May he . . . deign to govern, guard, and increase it, to the restoration of astronomic truth" (Rosen 1959, p. 131; Hugonnard-Roche 1982, p. 52). By the end of *Narratio Prima,* astronomical truth has been redefined; hypotheses have become what they are throughout *De Revolutionibus*: physical truths about the universe.

The Need for Rational Conversion in Scientific Revolutions

To explain his master's espousal of heliocentrism and to justify his own adherence, Rheticus invented rational conversion. He did so because the Copernican argument field was incomplete, inadequate in its own terms. Until Brahe, Kepler, and Newton, the heliocentric view fell short of its own highest goal: "to explain completely," as Copernicus says in the *Commentariolus,* "the structure and motions of the universe" (Rosen 1959, p. 90; my translation).[14]

To Copernicus, and therefore to Rheticus, "to explain completely" had at least two meanings. First, it meant a thoroughgoing realism: astronomers must describe the unique, that is, the actual, lunar and planetary orbits. On this reading, the Copernican revolution consisted not in heliocentricity only, but also, and perhaps primarily, in this insistent realism: "In this sense—if in no other—Copernicus may be considered the first great figure of the Scientific Revolution. It was essentially his attitude which came to prevail with Kepler, Galileo, Descartes, and Newton" (Grant 1962, pp. 215–216). Although *De Revolutionibus* was preceded by a preface (later shown to be written by Osiander) disavowing this view, no one studying the work with any care could have been deceived; certainly, the Church was not. In 1620 it approved the work only when it was appropriately emended: when Copernicus spoke of *demonstrating* the triple motion of the earth, the Church insisted that he speak only of the *hypothesis* of that motion (De Morgan 1954, p. 95).

But Copernicus could not fulfill his realist program. His epicycles and eccentrics—intricate combinations of uniform circular motions—could never describe unique orbits. This was not because epicycles and eccentrics could not be real; indeed, Averroes is criticized by Rheticus for his conclusion "that epicycles and eccentrics could not possibly exist in the realm of nature" (Rosen 1959, p. 194; Hugonnard-Roche 1982, p. 86). Instead, the problem was that epicycles and eccentrics combined to create too many solutions to lunar and planetary orbits. All were mathematically equivalent; consequently all had an equal claim on belief.

Copernicus was well aware of this equivalence (Rosen 1959, p. 74, note; see also pp. 138 and 168). Only one of these solutions could be correct; yet it was impossible to choose the actual among them—if any among them *was* actual. But if a real choice was impossible, then the Copernican model for the universe was, in this respect at least, a mere calculating device: "What has been determined cannot have innumerable explanations; just as, if a circumference is drawn through three given points not on a straight line, we cannot draw another circumference greater or smaller than the one first drawn"

(Copernicus 1971b, pp. 100–101). Although heliocentricity was real for Copernicus, and epicycles and eccentrics could be real, *his* epicycles and eccentrics remained mathematical conveniences. A passage from *De Revolutionibus* makes clear Copernicus' mature view in this matter, a blend of skepticism and hope: "Since so many arrangements lead to the same result, I would not readily say which one is real, except that the perpetual agreement of the computations and the phenomena compels the belief that it is one of them" (1978, p. 164).

To Copernicus, and therefore to Rheticus, to explain completely was also to explain simply. In *Narratio Prima,* early in his defense of heliocentricity, Rheticus entertains parsimony as an explanatory value: "This one motion of the earth satisfies an almost infinite number of appearances." Later in his argument Rheticus goes further: he assumes parsimony as an explanatory value: "With so few motions was it feasible to satisfy most of the necessary phenomena of nature" (Rosen 1959, pp. 137, 161; see also pp. 137 and 149).[15] But Copernicus' world system remains as he left it—a scheme far from simple, one that is no simpler than its Ptolemaic rival. True, as Copernicus illustrates it in *De Revolutionibus,* his universe seems simple enough, a system of seven nested spheres (an eighth forms the moon's sphere [Armitage 1962, p. 82]). But this is not an accurate picture: there are far too few circles. The drawing derives not from Copernicus' actual system but from a class of world-system depictions of which Peter Apian's much-reprinted geocentric universe is a typical example (Armitage 1962, p. 80). In Copernicus' illustration the earlier drawing is redrawn, with the positions of the sun and the earth simply reversed: "We are thus confronted" not with depictions, but with "ideograms [*Ideogramme*], of certain ideas and certain meanings. It involves a kind of comprehending where the meaning is represented as a property of the object illustrated" (Fleck 1979, p. 137). The diagram represents an act of faith in the physical possibility of a heliocentric system; nevertheless, it firmly "adhere[s] to Ptolemy's assumption of contiguous nested [solid] spheres" (Swerdlow 1976, p. 129).

If to explain completely meant to provide a unique solution to lunar and planetary orbits, one which described them simply and accurately as they actually were, Copernicus fell short of his announced purpose, not only in historical retrospect but in his own terms. For Rheticus, and for the early Copernicans, therefore, the change to heliocentricity was necessarily a movement of the will as well as the reason. Rheticus' invention of rational conversion to justify change acknowledged the obvious truth; in *Narratio Prima*, form followed function.

The Copernican Revolution as Rhetoric

When we see in *Narratio Prima* just how far short of his goals Copernicus fell, we may be tempted to argue against the very notion of a Copernican revolution, a position shared by I. Bernard Cohen (1985) and Arthur Koestler (1968). Their reasons are essentially the same: "The notion that a revolution in astronomy attended the publication of Copernicus' *De Revolutionibus* in 1543 was a fanciful invention of eighteenth-century historians of astronomy" (Cohen 1985, pp. x, 106; see also pp. 40–125). But Cohen and Koestler might have reversed their judgments of *De Revolutionibus* had they given more emphasis to the force of rhetoric in shaping the history of astronomy.

The evidence is certainly solid concerning the lack of immediate impact for Copernicanism: sixty-six years separate Copernicus' *De Revolutionibus* from Kepler's *Astronomia Nova,* the work in which, by announcing his first two planetary laws, Kepler inaugurates a new age of mathematical astronomy (Berry 1961, pp. 179 ff.). During those intervening years, the influence of Copernicus on most professional astronomers was clearly negligible. Indeed, Copernicanism was not accepted in England until 1650, or in Sweden, Spain, Hungary, or Poland until nearly the middle of the eighteenth century (*Colloquia Copernicana, I,* 1972). Although this opposition was often irrational, it could be and often was understandable and defensible, since the arguments for heliocentricity remained less than compelling. On this view, by the time Copernicanism was generally accepted, it is better described as a result of an evolutionary change.

But this view overlooks important evidence. Koestler calls *De Revolutionibus* "the book that nobody read" (1968, p. 191; emphasis omitted); to Cohen, Copernicus' masterpiece "had no fundamental impact on astronomy until after 1609" (1985, p. 38). This scarcely seems fair to a work that Rheticus devoured, Brahe knew well, Kepler studied assiduously, and Maestlin annotated over a period of fifty years (Moesgaard 1972; Kepler 1981, p. 17; Westman 1975a). These men formed a network of influence. Maestlin was Kepler's teacher, and Brahe his mentor; it was, in fact, Brahe who held the empirical key to the laws of planetary motion. When Kepler's *Mysterium Cosmographicum* appeared in 1596, it contained two contributions by Maestlin: a précis of Copernican astronomy and an introduction to a work appended to Kepler's: Rheticus' *Narratio Prima*. These connections belie the view that there was no early Copernican impact on astronomy; instead, they support the contention of Robert Westman that in the 1570s "a small, but significant group of mathematical astronomers had familiarized themselves profoundly with the technical aspects of Copernican astronomy and granted a serious hearing to its realist claims" (1975a, p. 54).

To trace this network of influence is to chronicle a progress that takes us from the birth of modern heliocentricity to the threshold of its first significant intellectual flowering: Kepler's three laws of planetary motion. Rhetorical links form a chain of descent from Rheticus to Maestlin, Kepler's teacher, and from Rheticus and Maestlin to Kepler himself, a chain, in fact, of rational conversions. Maestlin, who refers to Copernicus as "the Prince of Astronomers after Ptolemy," supports his commitment to the Copernican hypothesis as follows: "I do not wish to approve it on the grounds that I have been deceived and fascinated by love of novelty, but rather, compelled by necessity, I came to it with reluctance" (Westman 1972, pp. 9, 17). This statement has a familiar ring. Indeed, it paraphrases a passage from *Narratio Prima* (Rosen 1959, p. 186; see p. 102 of the present volume). But the influence of Rheticus on Maestlin is also readily apparent from the testimony in his appendix to the *Mysterium Cosmographicum* (1981, pp. 83–85).

Kepler is another link in the chain. The prefatory material to his *Mysterium Cosmographicum*, his eccentric first masterpiece, clearly documents Maestlin's influence as a Copernican teacher—"the distinguished Michael Maestlin at Tübingen." More to the point, it plainly attests to Kepler's rational conversion. First, he was "disturbed by the many difficulties of the usual conception of the universe" and "collected together . . . the advantages which Copernicus has mathematically over Ptolemy." Next, to prepare for a new teaching position, he was forced into "further contemplation." Finally, he experienced "a strong desire . . . [and] threw [himself] with the whole force of [his] mind into this subject [*toto animi impetu hanc materiam incubui*]"

(1981, pp. 62–63). Although the translation of the last clause is adequate, *animus* may refer also to the soul as the seat of the will, and *incubo* can also mean "to pass the night in a temple to receive a divine message" (*Cassell's*). Such linguistic inferences only strengthen an already strong case: Kepler explicitly calls this autobiographical process a complete conversion, which, nevertheless, was not undergone rashly (1981, pp. 78–79).

Directly and indirectly, then, Rheticus provided Kepler with the model for his rational conversion, his shift from one argument field to another, although sufficient "scientific" reasons were wanting. This rhetorical network, stretching over three generations, confirms the Copernican moment as revolutionary.

The need for rational conversion in heliocentric astronomy eventually vanished. This disappearance contributed to the illusion of objectivity, the illusion that the development of heliocentric astronomy did not depend on the conscious engagement of individual wills. But rhetorical analyisis teaches us otherwise. When Copernicanism began in the 1540s, the arguments in its favor were supplemented by the dramatic evidence of Rheticus' personal commitment. In the first century of the Copernican venture, such individual testimony and personal commitment continued in importance. By the end of the seventeenth century, with the Copernican revolution nearly over in the West, individual testimony and personal commitment gave way to their institutional surrogates, authority and indoctrination.

Rational conversion to heliocentric astronomy was more than a psychological bridge between fields of argument. The conflict between Copernicus' uncompromising realism and the deficiencies of the actual system he developed had a distinctly positive force: it acted as a century-long heuristic, inspiring a line of scientists that stretched from Brahe to Newton. Drawing their inspiration from this source, these scientists worked to create a universe they judged structurally and dynamically adequate: a mathematically simple and harmonious total system, in accord with the laws of a new physics and precisely descriptive and predictive of celestial appearances, a system that, in their view, accurately described the actual motions of real celestial objects. It was out of the efforts of Copernicus' successors, then, that there emerged a heliocentric universe, far different from that of Copernicus, but one more nearly in conformity with the explanatory ideal first asserted publicly in *Narratio Prima*.[16]

The role of rational conversion in scientific revolutions is not limited to the Copernican. In 1944 Oswald T. Avery and his collaborators published a paper containing a revolutionary conclusion: that genes were made of DNA. Concerning this discovery, Nobel laureate Salvador Luria reflects:

> No pressure or competition or sudden emergence of irrefutable data precipitated the decision to publish. In retrospect it appears that at this point Avery may have decided to exert the prerogative of the experienced leader. He exhibited the assertiveness that comes from the habit of success: a willingness to impose on a still confused mass of data a certainty that is emotional as well as rational. Such a source of certainty in science is unrecognized by those who believe certainty comes only after innumerable controls and attempts to disprove. The certainty Avery exhibited is more akin to illumination, a sudden vision projecting the possibility of an intellectual leap. (1986, pp. 29–30)

Later, Luria makes explicit the connection between the revolution in biology and that in physics: Avery published his results as "a *fact*— even though a fact validated as much by conviction as by hard evidence, a fact less fully convincing to experimentally minded critics than Kepler's laws might have been to Huygens and Newton."

CHAPTER 8

Newton's Rhetorical Conversion

In science there are two sorts of rhetorical masterpieces: those powerful enough to provoke revolution, and those ingenious enough to avoid it. *Dialogue Concerning the Two Chief World Systems* and *On the Origin of Species* are examples of masterpieces of the first sort. Initially, Galileo and Darwin caused more debate than assent, more turmoil than change. Descartes's optical works and Newton's *Opticks,* on the other hand, are masterpieces of the second sort: each in its own way successfully persuaded; each dominated optical research for nearly a century. In these works, each man argued for change less as revolution than as continuity: the extension of the best of the past. Despite the similar persuasive aims of these masterpieces, however, their notions of science could not be more opposed.

It is this opposition that is my subject, a contrast that reaches to the heart of seventeenth-century optics. Published in 1637 as an integral, though now nearly forgotten, appendix to his *Discourse on Method,* Descartes's optical works have quite properly been called a first step in a new direction. Although Descartes did not violate traditional presuppositions, he created from these an original physics of light. Despite the newness of his program, he shared with traditional science the conviction that white light was basic while color was derivative, an alteration of white light. He also believed with tradition that rational intuition, not experiment, was epistemologically prior: reason, not experience, was the bedrock, the touchstone, of knowledge. Finally, he was as convinced as Aristotle that a complete scientific explanation must include at least the traditional three causes: the formal, the efficient, and the material. These shared convictions allowed Descartes to be rhetorically transparent and convincing at the same time: he could use his prose to clarify, indeed to highlight, his views.

In 1672, a time when Descartes's optics was already firmly entrenched, Newton published his first paper on that science. Its views were new in a new way. Unlike Descartes, Newton challenged a traditional tenet concerning the nature of light. For the first time in the history of optics, white light was revealed as derivative, as a compound of all of the lights of the visible spectrum. Moreover, the ground Newton offered for certainty in optics differed radically from that of his predecessor, and of his predecessor's predecessors: by giving epistemological priority to experiment over rational intuition, Newton overturned a central presupposition of traditional science. In addition, Newton's explanation of light did not give a full account of its origin, an account that included the operation of its material cause. A startling claim, a new method, a different, more restrictive, style of explanation: seemingly, Newton needed to discharge a strong burden of proof. But in this early paper on light and color, in a rhetoric as transparent as that of Descartes, Newton did not discharge this burden; instead, he emphasized his conflict with traditional views and methods.

After a flurry of inconclusive debate, and a second early paper, he withheld his optical researches from the public for nearly three decades. In 1704 he published his *Opticks,* his second attempt at persuasion. In this work Newton discarded the transparent rhetoric that made the epistemological and explanatory novelty of his early papers clear; in its stead, he substituted a rhetoric that invented an essential continuity between his work and the optical and scientific past. The rhetoric of the *Opticks* concealed his radical intent; it was designed to convince at the expense of frankness. In his final masterpiece Newton transformed optics, and experimental science, by allowing his readers to believe that an adherence to the new did not entail a fundamental rejection of the old. This strategy was successful: throughout the eighteenth century, in England and on the Continent, the physics of light was Newton's physics.[1]

The Background of Cartesian Optics

An argument field is a set of "self-evident" field-specific truths and their inference rules. From such truths, by means of these rules, conclusions are drawn: to secure a claim rationally is "[to use] a rule of inference vouched for by agreeable authorities" and "[to violate] no taken-for-granted assumptions" (Willard 1983, p. 91). From field to field, rules of inference differ in pattern and rigor: in number theory and in rhetorical theory, for example, such rules will probably continue to differ in these important ways. Rules can also vary within the same field over time: for instance, the seventeenth-century development of the calculus depended importantly on a relaxation in the rigor demanded of mathematical proofs. Finally, it is a mistake to feel that all of these inferential procedures can be reduced to a single set: although mathematics and logic may share an equal rigor, the former cannot be reduced to the latter. Indeed, as Schuster makes clear, the methodological rules for doing science cannot be reduced to a single set.

The central truths and agreed-upon methods of traditional optics constitute an argument field in science. All are present in Descartes. He agreed that white light was basic, and color derivative, an alteration in white light. He also concurred with tradition in representing light as straight lines that can cross without interfering. This characteristic permitted him to use geometry to solve problems in the physics of light: the axioms and theorems of Euclidian geometry are central to the methodology of both traditional and Cartesian optics.

To illustrate the two central geometric properties of sunlight— its rectilinear propagation and the noninterference of its rays— Descartes compares light's action to the pressure on a vat of half-pressed grapes. This pressure transmits itself throughout the vat, squeezing the juice equally through two widely separated holes in the bottom:

> And in the same way considering that it is not so much the movement as the action of luminous bodies that must be taken for their light, you must judge that the rays of this light are nothing else but the lines along which this action tends. So that there is an infinity of such rays which come from all points of luminous bodies, toward all points of those that they illuminate, in such a manner that you can imagine an infinity of straight lines, along which the actions coming from all points of the surface of the wine . . . tend toward [one hole], and another infinity, along which the actions coming from these same points tend also toward [the other hole], without either impeding the other.[2]

When geometric methods are applied, light is found to exhibit two regularities: reflection and refraction. When a ray of light is reflected, its angle of incidence, the angle at which it strikes the reflecting surface, is always equal to the angle of reflection, the angle at which the ray leaves the reflecting surface (Figure 5). When light penetrates a transparent body at an angle other than right, it refracts; it bends toward or away from an imaginary line, called the normal, a line perpendicular to the refracting surface. When the light crosses into a medium of higher optical density, for example from air into water, it bends toward the normal; when it crosses into a medium of lower optical density, it bends away (Figure 6).

The angles of incidence and reflection are equal, a quantitative relationship known since antiquity. An analogous quantitative relationship for refraction eluded all investigators until Snell and Descartes. In refraction, as both discovered, rays of light conform to a definite geometric relationship, that of sines. In a triangle one of whose angles is right, the sine is the ratio of the side opposite either acute angle and the side opposite the right angle (Figure 6, inset). A ray refracting in most transparent substances bends in such a way that the ratio of the sine of its angle of incidence to the sine of its angle of refraction equals a specific number. For each regularity-abiding refracting substance, this number never changes:

$$\frac{\text{sine } i}{\text{sine } r} = \text{constant}$$

Although the sine relationship was a new discovery, it was an empirical regularity compatible with traditional assumptions and methods: there was no doubt concerning its immediate absorption into the body of traditional optics.

The New Cartesian Science

Although Descartes practiced traditional geometric optics, he did so within the framework of the new science he created: Cartesian optics transforms traditional optics into a new argument field. Scientists working within traditional optics might easily be persuaded that the refractive regularity noted by Snell and Descartes was not merely a useful empirical discovery but a law of science—wrongly persuaded, in Descartes's view. He held that an empirical regularity becomes a

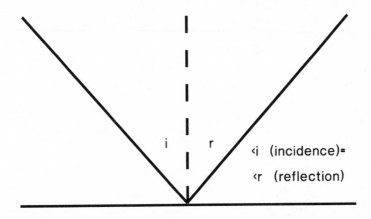

Figure 5   Reflection.

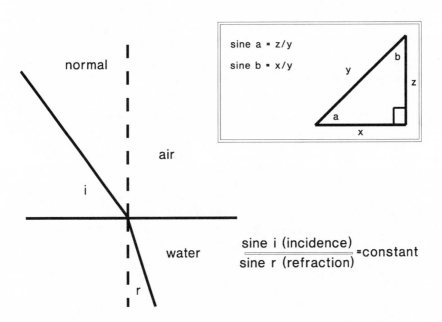

Figure 6   Refraction.

scientific law only when it is understood as an integral part of a true and coherent physics, his physics. Thus the refractive regularity of Snell and Descartes can be promoted to a law only when we understand it as correct according to the central and absolutely certain principles of Cartesian physics.

Descartes shares with tradition his conviction that the goal of his science is absolute certainty rather than moral certainty (what we would call very high probability). But the source of his certainty represents a radical interpretation of the primacy of reason in epistemology, one that leads to a new, counterintuitive physics. Provided they are clear and distinct, the intuitions of Descartes's unaided reason form the incorrigible foundation of his science.[3] According to these, the material world consists only of extension and its laws of motion. The universe is a plenum, completely full: space itself is material, because extended. Local motion, the relative repositioning of extended substances, can occur only through contact. Everything we experience, even light, is the result of local motion originating in contact.

In Cartesian physics, the sun consists of small, rapidly moving particles of the first element which exert a pressure on the second, the element that fills interplanetary spaces. This pressure is transmitted instantaneously in straight lines in all directions. Experienced on the retina, this pressure creates the sensation called light. For Descartes, light *is* this pressure; it is not, as it was for Aristotle, a qualitative alteration in a medium, the air; it is not, as it will be for Newton, the effect on the retina of a stream of tiny particles traveling from a luminous source.

At the end of his *Principles of Philosophy* Descartes sums up the full implications of his position:

> These reasonings of ours will perhaps be included among the number of these absolutely certain things by those who consider how they have been deduced in a continuous series from the first and simplest principles of human knowledge. Especially if they sufficiently understand that we can feel no external objects unless some local movement is excited by them in our nerves; and that such movement cannot be excited by the fixed stars, very far distant from here, unless some movement also occurs in these and the whole intermediate heaven; for once these things have been accepted, it will scarcely seem possible for all the rest, at least the more general things which I have written about

the World and the Earth, to be understood otherwise than as I have
explained them. (1983/84, pp. 287–88)

In Descartes's physics, then, reason is sole guarantor of the certainty of the general principles from which the specific truths of this
science are derived. In a typical commment on the relationship
between reason and experiment, Descartes asserts: "And the demonstrations of this [law of impact] are so certain that, even if experience were to appear to show us the opposite, we would nevertheless
be obliged to place more trust in our reason than in our senses"
(1983/84, p. 69n.). It is our reason alone that assures us: no explanation of light, no explanation in physics, can be persuasive if it is
not also mechanical, the result, direct or indirect, of matter in motion:
"Whatever I concluded to be possible from the principles of my
[natural] philosophy actually happens whenever the appropriate
agents are applied to the appropriate matter" (1970, p. 38).

Like light, color must be explained in terms of matter in motion.
In Descartes's view, we recall, the particles of the second element,
those that fill all the space between the sun and the earth, transmit
the sun's first-element pressure. In refraction, these second-element
particles take on differing rotational speeds, speeds whose pressures
are communicated directly to the retina. It is from these differing
pressures that the sensations of color arise: "The nature of the colors
appearing at [the lower end of the spectrum] consists only in the fact
that the particles of the fine substance that transmits the action of
the light have a stronger tendency to rotate than to move in a straight
line; so that those which have a much stronger tendency to rotate
cause the color red" (1965, p. 337). For Descartes, the sensation of
white light is the result of the simple pressure of particles of the
second element on the retina; color is the result of the action of these
same particles to which a rotating motion has been imparted.

Although reason is central to Cartesian physics, experience cannot
be ignored; indeed, observation and experimentation have an important role.[4] Though everything flows from general principles, everything cannot be deduced from them: the constant of optical density,
which accounts for the degree of refraction, varies for different
substances and cannot be deduced from the law of refraction. To
obtain this constant in each particular case, "we must appeal to experience" (1965, p. 81). Even the law of refraction, though it conforms

to the laws of motion, cannot be deduced from them. Light always acts in accordance with the laws of motion, which are general principles, but it does not always act in accordance with the law of refraction. Experience is the sole desideratum for empirical regularities: "For, seeing that these parts could have been regulated by God in an infinity of diverse ways; experience alone should teach us which of all these ways He chose" (1983/84, p. 106).

An examination of Descartes's philosophy, it is true, reveals a role for observation and experiment that is more nearly central. In that philosophy, reason tells us that the real qualities of material objects with which science is concerned consist "clearly and distinctly" of "magnitude or extension in length, breadth, or depth" (1931, I, 164; see also pp. 56 ff. and 1983/84, pp. 76–77). Such a view seems to entail a quantitative physics. But neither mathematics nor measurement appears regularly in Descartes's science; his works contain pages filled with accounts not susceptible to measurement, accounts no experiment could ever confirm.[5] Despite his statements to the contrary, Descartes's physics seems essentialist and qualitative (cf. Gaukroger 1980, p. 134).

In all of Descartes, there is one exception to this subordinate epistemological role for observation, an exception that proves this rule. In his discussion of the rainbow, Descartes signally advances the cause not only of his physics but of ours; by means of observation and measurement he deduces the radii of both rainbow arcs: "the radius of the interior arc must not be greater than 42°, nor that of the exterior one smaller than 51°" (1965, p. 339). *Must be!* From Descartes's discourse on the rainbow, which Newton knew well (1969, III, 543–549), one could learn a whole new way of doing science, a way in which observation and measurement combined to form an epistemological vanguard in search of new, durable empirical regularities.

But from this analysis, which he regarded as an authentic example of his method (1970, p. 46), Descartes learned nothing new about the fundamental role of observation and measurement: he gives equal credence to his theory of the arc of the bow, amply confirmed by observation and measurement, and to his theory of color, for which he offers, in addition to a general conformity with experience, only the certainty of his unique, reason-derived mechanics (1965, p.

338). Concerning this latter certainty, Descartes sounds a characteristic note when he asserts "how little faith we must have in observations which are not accompanied by true reason" (1965, p. 342).

Newton's First Paper on Optics

In his first paper on optics, published in the *Philosophical Transactions* of 1672, Newton tells the story of his discovery:

> In the beginning of the Year 1666 . . . I procured me a Triangular glass-Prisme, to try therewith the celebrated *Phaenomena* of *Colours*. And in order thereto having darkened my chamber, and made a small hole in my window-shuts, to let in a convenient quantity of the Suns light, I placed my Prisme at his entrance, that it might be thereby refracted to the opposite wall. It was at first a very pleasing divertisement, to view the vivid and intense colours produced thereby; but after a while applying my self to consider them more circumspectly, I became surprised to see them in an *oblong* form; which, according to the received laws of Refraction, I expected should have been *circular*.
> (1978, pp. 47–48)

As Newton tells his story, the anomaly of the oblong spectrum demands an explanation: if light behaved as Descartes said, the spectrum would disperse equally in all directions, like spray from a hose. At first Newton attempts explanations without any, or without serious, theoretical import: perhaps it was the thickness of the prism-glass, perhaps its unevenness; perhaps the rays curve. All of these "suspicions" being removed, Newton performs the "*Experimentum Crucis,*" the decisive experiment; the final searching question is put to nature (1978, p. 50).

To this question the optical tradition has no answer: it exhibits a fatal flaw, a deep-seated but hitherto unsuspected incoherence. Unless we abandon the law of sines—an unthinkable prospect—we must abandon the belief that white light is basic, color derivative; we must believe instead that color is basic, white light derivative. The crucial experiment clearly and unequivocally shows that "[white] *Light* consists of *Rays differently refrangible* [bendable]" (Newton 1978, p. 51). When these rays are separated, they form the radiant colors of the spectrum; combined again, in the proper proportions, they recreate white light.

In thirteen propositions that follow the *Experimentum Crucis*, Newton shows how the differing refrangibilities of light explain "the celebrated *Phaenomena* of *Colours*" (1978, p. 47). Descartes's views notwithstanding, sunlight is "ever compounded, and to its composition are requisite all the aforesaid primary Colours, mixed in a due proportion" (1978, p. 55). It follows from the combinatory nature of white light that "Colours are not *Qualifications of Light,* derived from Refractions, or Reflections of natural Bodies [as in Descartes]. . . but *Original* and *connate properties,* which in diverse Rays are divers" (1978, p. 53). This difformity of light explains not only radiant colors (those produced by a prism) but "the Colours of all natural Bodies [which] have no other origin than this, that they are variously qualified to reflect one sort of light in greater plenty then another" (1978, p. 56).

It is a deliberate irony that the spearhead of Newton's attack on Descartes is the certainty of an anomaly founded on experience: it will not go away; it cannot be explained away. It is equally ironic that only a wholly quantitative point of view can reveal the anomaly: unlike Descartes, Newton projects his spectrum not just a few inches, where the anomaly is not apparent, but 22 feet, where it is unmistakable (Westfall 1984, p. 164). Like Kepler's famous 8 minutes of arc, these 22 feet have revolutionary implications in Newton's hands, and in his early paper he makes the most of them.

In this paper, following in the footsteps of Boyle and Hooke, Newton uses experiment systematically as the primary epistemological instrument. But Newton's relentless use represents a new departure, as a comparison with *Experiments and Considerations Concerning Colors* and with *Micrographia* will confirm. In theory and in practice, Newton clearly and uncompromisingly reverses the traditional and Cartesian roles of reason and experiment: "If the Experiments, which I urge, be defective, it cannot be difficult to show the defects; but if valid, then by proving the Theory they must render all Objections invalid" (1978, p. 94). In Newton's science, experiment bears the full weight both of discovery and of theory.

In a justly famous Baconian pronouncement on method, Newton affirms experimentation as the driving force behind theory, and the prime source of certainty. In the same passage he discloses his perception of the central defect of Cartesian physics, indeed of any physics that overemphasizes the hypothetical—the fatal ease with which explanations may be invented:

> The best and safest way of philosophizing [doing science] seems to be this: first to search carefully for the properties of things, establishing them by experiments, and then more warily to assert any explanatory hypotheses. For hypotheses should be fitted to the properties which call for explanation, and not be made use of for determining them, except in so far as they can furnish experiments. And if anyone makes a guess at the truth of things by starting from the mere possibility of hypotheses, [I] do not see how to determine any certainty in any science; if indeed it be permissible to think up more and more hypotheses, which will be seen to raise new difficulties.[6]

Newton's first paper failed to convince. His notion of a crucial experiment, one whose result is proof against argument, is both central to his case and seriously flawed as a persuasive device. The persuasiveness of the crucial experiment depends on its replicability; but the crucial experiment in this first paper is accompanied by neither a diagram nor clear directions. More significant, because far less remediable, is the fact that the persuasive effect even of replicable results depends absolutely on their univocal interpretation. At one point Newton becomes exasperated with some rival experimental results: "It is not the number of Experiments," he cries, "but weight to be regarded; and where one will do, what need many?" (1978, p. 174). But it was folly to expect such unanimity of interpretation in so hotly disputed a seventeenth-century subject as the nature and action of light. In his first paper Newton draws what for him is the obvious ontological inference: "Since Colours are the *qualities* of Light, having its Rays for their intire and immediate subject, how can we think those Rays *qualities* also, unless one quality may be the subject of and sustain another; which in effect is to call it *Substance*."[7] But Hooke and Pardies had different, though plausible, explanations of Newton's experimental result, explanations founded on the rival ontology of wave theory.

Critics also objected to the incompleteness of Newton's explanations. Huygens animadverted that "he hath not taught us what it is wherein consists the nature and difference of Colours, but only this accident . . . of their *different Refrangibility*" (1978, p. 136). Indeed, Newton had avoided any mechanical explanation of refraction or color formation: "I shall not mingle conjectures with certainties," he grandly asserted (1978, p. 57). But for Huygens, a scientific explanation that insisted on the corpuscularity of light must include a specific mechanism for the production of the sensations of light and color by means of matter in motion.

Newton's 1672 paper succeeded in establishing neither his method of doing science nor his beliefs about the nature and action of light. In his *Opticks* of 1704, Newton created a second opportunity to persuade the scientific community of the efficacy of his method and the truth of his optical beliefs—the same method and essentially the same beliefs contained in his optical papers of thirty years before.

Newton's *Opticks:* A Rhetorical Masterpiece

The rhetoric of Newton's first paper—riveting to this day—was that of youth: brash and brilliant, relying for its persuasive effect on a clash of principles, the paper was a decisive confrontation, clearly and unequivocally presented. In contrast to his early papers, the rhetoric of the *Opticks* was that of late middle age: a canny and successful attempt to transform a youthful invention into a durable inheritance. In the exposition of his theory in the *Opticks*, Newton employed a Euclidean arrangement to create an impression of historical continuity and logical inevitability.[8] In addition, by piling experiment on experiment, and, in each experiment, detail on detail, he created in this work an overwhelming presence for his experimental method. Finally, in the book's last section, he initiated a cascade of rhetorical questions whose cumulative effect was both to sanction his science and license his speculations.

The Use of Arrangement

"Let me suggest," says David Lindberg,

> that optics from Alhazen to Newton (and, in some respects, from Aristotle and Ptolemy to Newton) must be seen as a continuously unfolding discipline. Historians of science have become overly conditioned to perceive the sixteenth and seventeenth centuries as a revolutionary period, during which scientists made a clean break with the past. There is, of course, much truth in this view, even in optics; but if pushed too far, it obscures the large measure of continuity in the history of optics from Alhazen onward. (1968, p. 36)

Lindberg's evolutionary interpretation of optical history has a New-tonian parallel. In February of 1676, in a letter to Hooke, Newton explicitly acknowledges his place in a long optical and scientific tra-dition: "What Des-Cartes did [the discovery of the sine law] was a good step . . . If I have seen further it is by standing on ye sholders of Giants" (1959, I, 416).

There is a vast difference in intent between the historian and the scientist. Lindberg presents the essential continuity of optics as a historical interpretation, an end; but for Newton, in this early letter, historical continuity is a means, the outline of a persuasive strategy. This strategy—so different from that of his early paper—is fully embodied nearly three decades later in his *Opticks*.

In Newton's early paper, the dominant arrangement is narrative; in his *Opticks*, virtually the same material is set out as a Euclidean deduction. In the early paper, Newton uses narrative to dramatize the clash between past and present; in the *Opticks*, he uses Euclid to display the present as a deductive consequence of the past: "I have now given in Axioms and their Explications the sum of what hath hitherto been treated of in Opticks. For what hath been generally agreed on I content my self to assume under the notion of Principles, in order to what I have farther to write" (1952, pp. 19–20).

In the *Opticks*, however, there is a tension between the deductive form Newton invariably adopts and the inductive epistemology he routinely employs. In the crucial first book, Euclidean form is partic-ularly strict: definitions and axioms are followed by a series of prop-ositions with their proofs. But the method of all but the last Propo-sition is that of analysis, induction rather than deduction; it "consists in making Experiments and Observations, and in drawing general Conclusions from them by Induction, and admitting no Objections against the Conclusions, but such as are taken from Experiments, or other certain Truths" (1952, p. 404). Only in the last Proposition does the method change to synthesis or "Composition": the nature of light and of both radiant and natural color having been proved "by Experiments," they are now demonstrated by deduction (1952, pp. 405 and 20).

That this blurring is deliberate is clear from Newton's suppression of the explicit epistemological introduction to the first edition (West-fall 1984, pp. 640–644). To move from the early papers to the *Opticks* is not to move from one science to another but from one rhetoric to another—to pass from a work that continually clarifies the epistemological priority of experiment to one that obscures the distinction between the traditional view of experiment as confirmatory and Newton's more radical view. For the initiated, the emphasis on experiment underscores its radical epistemological role; but this emphasis can bear a more conservative interpretation: it allows Malebranche to say that Newton's theory "fits (*s'ajuste*) . . . all his experiments" (Guerlac 1981, p. 110). In the *Opticks*, Newton deliberately includes in his scientific audience even those who might differ from him on epistemological grounds.

The Use of Presence

The epistemological ambiguity of the experiments described in the *Opticks* contrasts vividly with their clear and complete presentation. Throughout, the presence of Newton's experimental method is enhanced by the sheer number described, and by the quantitative meticulousness with which their methods and results are reported. For any who might find experiments persuasive, Newton describes not two, as in his first paper, but dozens, all set out with a meticulous detail arguably new to science: "I have set down such Circumstances, by which either the Phaenomenon might be render'd more conspicuous, or a Novice might more easily try them" (1952, p. 25). In Newton's first paper even the *Experimentum Crucis* was meant to have its effect—an effect nothing less than the overturning of traditional optics—without the benefit of either a diagram or sufficient detail for easy replication. In contrast, in the *Opticks*, a typical experiment is accompanied by a detailed diagram and a description that begins thus: "In the middle of two thin Boards I made round holes a third part of an Inch in diameter, and in the Window-shut a much broader hole being made to let into my darkned Chamber a large Beam of the Sun's Light" (1952, p. 45).

This concern for the persuasive value of meticulous detail extends to measurement. Everywhere in the *Opticks*, measurement, so important in the early papers, increases in importance; in seemingly every case, measurements previously made are remade: in book II, part II, of the *Opticks*, for example, Newton recalculates the values of a table he published in the *Transactions* nearly thirty years earlier (1978, p. 219; 1952, p. 233). The differences are sometimes very small indeed. In the earlier work the thickness of glass at which indigo of the second order is most intense is .0000085; in the *Opticks* the figure is .000008182, a difference of 318 *billionths* of an inch!

The Use of Rhetorical Questions

Halfway through the first part of the third book a remarkable event occurs: both the Euclidean structure and the experimental program break off completely, and a long section of Queries is initiated, each in the form of a negative rhetorical question.[9] The question marks that punctuate these Queries bestow upon Newton a double benefit: they underline their speculative nature, but leave undiminished their rhetorical force; by their very nature, negative rhetorical questions are strong positive assertions.[10] In the body of the *Opticks*, a light ray is defined as "the least Light or part of Light, which may be stopp'd alone without the rest of the Light, or propagated alone, or do or suffer any thing alone, which the rest of the Light doth not or suffers not" (1952, p. 2). The definition is operational: light behaves as if it were composed of rays made up of parts. But in the Queries, light rays are unequivocally physical entities: "Are not the Rays of Light very small Bodies emitted from shining Substances?" (1952, p. 370).

The dual character of the Queries permits Newton to meet some crucial objections of his early critics. Hooke and Pardies had criticized Newton's inference that light was a substance. In the *Opticks,* Newton blunts this criticism by a distinction. In the body of this work he confirms by experiment that light behaves consistently as if it were a substance; in his Queries he makes a plausible, but not a scientific, case that light is, in fact, a substance. Newton's distinction between science and speculation also blunts Huygens's complaint that Newton's explanation of light was incomplete because it did not include a specific mechanism for its propagation. In his *Opticks* Newton shows his willingness to speculate about mechanisms, so long as speculations are not confused with science. In the Queries, for example, Newton considers the anomalous refractive behavior of light when passed through island crystal. From this behavior he infers a physical property of rays of light, one that might account for the anomaly: "Every Ray may be consider'd as having four Sides or Quarters, two of which opposite to one another incline the Ray to be refracted after the unusual manner" (1952, p. 360). But this mechanical explanation is speculative, not scientific.

In his "Account," Newton describes the Queries in the following way: "In this Philosophy Hypotheses have no place, unless as Conjectures or Questions proposed to be examined by Experiments. For this Reason Mr. *Newton* in his Optiques distinguished those things which were made certain by Experiments from those things which remained uncertain, and which he therefore proposed in the End of his Optiques in the Form of Queries" (Hall 1980, p. 312).

Despite their speculative character, the Queries serve certainty by staking out its limits. By clearly demarcating the opposing edges of science and speculation, they affirm the scientific status of the conclusions that precede them, as well as the methods by which these conclusions were reached. At the same time, the rhetorical forcefulness of the Queries, enhanced by the aging Newton's overwhelming reputation, clearly intend a constraint on the future of optical research. Newton sets his Queries down "in order to [facilitate] a farther search to be made by others" (1952, p. 339). In the body of the *Opticks,* the properties of light are experimentally demonstrated and deductively proved; in the Queries, the ground plan for the next century of optical research is successfully laid out.[11]

Solely by means of its rhetoric—by means of its strict Euclidean form, its striking experimental presence, its provocative speculations—Newton's masterpiece became the model for optics, and for experimental science, in the next century. Initially, scientists did not have to believe in the epistemological priority of experiment, or in the particle nature of light, to accept Newton's empirical results and to acknowledge his scientific leadership. But that acceptance and that acknowledgment eventually transformed optics into a modern experimental science along Newtonian lines; eventually, to do optics was to accept particle theory and to presuppose the epistemological priority of experiment.

The Problem of Paraphrasability

Each of Newton's two reconstructions of his optical past was an attempt to persuade his scientific audience of the truth of his unchanging views. In the early papers Newton emphasized the discontinuities between his theories and the optical past; he cast his work in a revolutionary mold. In contrast, in his *Opticks,* he emphasized the continuity between those theories and the past, presenting his work as an evolutionary development. Was Newton's optics revolutionary or evolutionary? The question is unanswerable. Historical continuity and discontinuity are not discovered; they are invented by rhetorical means to suit particular persuasive purposes. The lesson of Newton's and Descartes's optics is that there is no privileged reconstruction of the past; like the rest of us, scientists recreate their past to reflect the importance of the present they favor.

But even if one admits that the revolutionary or evolutionary character of Newton's optical theory is a rhetorical *trompe-l'oeil*, one may still insist that the theory itself represents a claim independent of its, or of any, rhetoric. Let us be as strict as possible; let us present scientific claims and arguments in logical form: "If, and only if, *a*, then, necessarily *b*; if, and only if, white light is a compound of all the spectral colors, each consisting of particles in mechanical interaction, then, necessarily, the solar spectrum will be oblong." One may quarrel with this particular schematic representation of Newton's central empirical regularity, and of the theory by which he explains it; but, in fact, any such representation will accomplish the same task: it will discard the style and alter the arrangement of the original only to replace them with a style and arrangement of another, more austere, sort. Style and arrangement in science are not veils that can be removed to reveal theory, the scientific core beneath.

One may concede as much. Nevertheless, one may point out that the scientific content of a text is precisely that which survives its legitimate paraphrases; there really is a scientific core, even if it can never actually be freed from all linguistic embodiments. But what is it to say that Newton's *Opticks* is the scientific equivalent of his early papers on light and colors? It is to say that to paraphrase either is to say the same scientific thing. Paraphrase, however, is not linguistic equivalence, the meaning that persists, for instance, through the active-passive transformation; it is, rather, a theory about a text, an equivalence claim whose support is the beliefs of particular audiences concerning science. We can say that Newton's optical papers and his *Opticks* share the same content only because we know what experiments and theories are, only because we can agree when we say that an experimental result verifies a theory, that a theory is entailed by an experimental result.

We may be inclined to say that the scientific content of texts can be paraphrased because they contain a core of meaning. In fact, just the reverse is the case: the core of meaning is what we paraphrase consistently. In the particular case of Newton's optics, we would regard as radically defective any paraphrase that omitted either his central empirical regularity or its explanation. But to include these is nothing more than to fulfill an expectation. Paraphrase is theory-driven; this theory selects the invariants we call the core of scientific texts. Our ability to paraphrase is only evidence that we have already been persuaded, that rhetoric has already done its work.

In their optical works, both Newton and Descartes invented their relationship to the past; indeed, both invented pasts appropriate to the presents they preferred. Newton differed only in that he invented *two* pasts, discarding the first when it proved inadequate to his persuasive purposes.

Both scientists also invented methods with which they hoped to make their scientific conclusions certain. Although there is no doubt that each believed in the efficacy of his respective method, Feyerabend makes clear "the essential vacuity of Newton's central methodological rule," and Schuster suggests that Descartes's famous method is similarly empty. For Schuster, this points to the essential vacuity of all such rules.[12] They are all mythic speech: none is true—there can be no method for all of science—though all are based on evidence that is incontrovertible to believers.

This conclusion in no way undermines the usefulness of method as rhetoric, "as a discourse partly constructed by ideological commitments or investments in particular scientific theories" (Schuster and Yeo 1986, p. xxiv). In their attitude toward scientific certainty, Descartes and Newton were deterred neither by their own doubts nor by the certain skepticism of future generations, a skepticism parallel to their own concerning the conclusions of past scientists. Descartes and Newton rightly emphasized not the tentativeness of their mistaken conclusions, but the correctness, the inevitability, of their results.

CHAPTER 9

Peer Review and Scientific Knowledge

A thorough examination of science from the point of view of persuasion cannot be confined only to its formal communications. Scientific books and journal articles are just the most visible products of verbal interaction within the community, posed photographs of a continuous activity captured at certain ritually significant moments, designated occasions suited to studio portraiture. We need to investigate not only the finished products but also the earlier stages of the complex persuasive process by which new science passes from private to public, from laboratory notebook to student textbook. In this chapter I will focus on one such stage, peer review, which represents the first step toward public status for scientific claims, toward the transformation of these claims into scientific knowledge.

To analyze peer review, I will use speech act theory and, especially, Jürgen Habermas's ideal speech situation, a construct derived from that theory. This coupling of theory and research site is justified by a coincidence of aims: rational consensus, the explicit *telos* of the ideal speech situation, is the avowed *telos* of peer review. If Habermas's theory is plausible at all, the application of its machinery must be plausible in this instance.

In speech act theory, the submission of a paper to a scientific journal counts as a request, a regulative act whose successful completion depends on shared social norms: communicative action is initiated and results, eventually, in a decision to accept or reject, a decision generally without appeal.[1] In all cases, a referee's report is an assessment of the persuasiveness of a submitted paper. If the judgment is negative, questions or comments will be regarded as support for a decision to reject. If the judgment is positive, these same questions or comments will be seen as instructions to the author concerning the paper's repair. When, as in the documents examined in this chapter, acceptance is conditional on repair, a communication network is created; it consists of referees, editors, and authors.

This set of peer review transactions can be analyzed in terms of the criteria of the ideal speech situation, which purports to be a set of suppositions unavoidable if we are to create rational consensus, to detect its presence, and to judge its quality. Perfect symmetry is required: each interlocutor must reveal himself, each must have every opportunity to initiate communication and to make assertions, and each must have equal control over the exchange; in the pursuit of rational consensus under these conditions, bias is revealed, countered, and neutralized. A peer review decision made under the conditions of the ideal speech situation represents a consensus among referees, editors, and authors that a particular paper is publishable science.

The sufficiency of the presuppositions of the ideal speech situation has been questioned because the theory does not "[refer] to the intelligence, competence, psychological normality, etc. of the participants," nor does it include such categories as "the nature (content) of cultural traditions" and "the distribution of material resources" (McCarthy 1973, p. 150; Held 1980, p. 396; see also Thompson 1982, p. 129). Clearly, in a complete account, the ideal speech situation must be sensitive to these sources of deviation from symmetry. But Habermas believes that this sensitivity is built into the theory at the formal level; furthermore, he sees the detection of any resulting deviation as essentially unproblematic: "Should one party make use of privileged access to weapons, wealth or standing, in order to *wring* agreement from another party through the prospect of sanctions or rewards, no one involved will be in doubt that the presuppositions of argumentation are no longer satisfied" (McCarthy 1973, pp. 150–151; Habermas 1982, pp. 272–273).

But there is a more central theoretical doubt. According to Habermas, the ideal speech situation is not generalization from experience; rather, it is a rational reconstruction of the criteria that make a certain kind of experience possible. Whether or not these criteria are realized in a particular communication, or indeed in any communication, is irrelevant to their theoretical status: together, they form "the constitutive conditions of rational speech" (1973 pp. 258–259). But even so sympathetic and knowledgeable a critic as Thomas

McCarthy doubts the intellectual viability of Habermas's version of Kant's transcendental deduction: "Rational reconstructions of universal or species competences cannot make the strong *a prioristic* claims of the Kantian project. They are advanced in a hypothetical attitude and must be checked and revised in the light of the data, which are gathered *a posteriori* from the actual performances and considered appraisals of competent subjects" (McCarthy 1982b, p. 60). Indeed, Habermas himself agrees that "linguistic reconstruction . . . calls for empirical inquiries undertaken with actual speakers" (McCarthy 1982b, p. 61).

This chapter is such an inquiry, an attempt to test the ideal speech situation against an actual case. I will focus on the contents of one issue of one biological journal, following the trail of those papers that initially generated substantial disagreements among the correspondents but were eventually accepted for publication. These disagreements created a rich collection of texts: full cycles of referee criticism, editorial mediation, authorial rebuttal and repair.

Deviations from the Ideal Speech Situation

The ideal speech situation permits each interlocutor an equal opportunity to initiate speech acts. In an obvious way, a review cycle deviates from this principle: after their initial request, authors never initiate speech acts but only respond to those of the referees or the editors. In addition, there is usually only one round of referee critique, editorial mediation, and author response, hardly enough to ensure the mutual understanding that must precede rational consensus. Moreover, because interaction between referee and author is entirely out of the question, authors are forced to respond without benefit of interactive clarification. They must reply to every question and respond to every criticism despite the fact that, were interaction possible, some questions might not have been asked, nor some criticisms made.

Habermas's second criterion is that all interlocutors have the opportunity to reveal their "inner natures," to allow their discourse to become transparent to their full subjectivity. Since referee reports and author responses are frequently full of the emotion-laden language routinely absent from scientific papers, we may infer that peer review exchanges fulfill this criterion. This inference would be mistaken. Ideally, peer review communication operates under two constraints: discourse must be free of emotion, and it must be polite. If these constraints are set aside, it is invariably referees who are the first offenders; only then can authors follow suit with impunity. For example, a referee carps:

> I have serious doubts about the numbers unless some moderately magnified, representative micrographs, of good quality, of the different tissues, are furnished. The reader is entitled, at least with the present state of the art and acceptance of stereology [the science of three dimensions] as a quantitative probe, to determine visually, if only intuitively, whether the numerical data may be reliable.

An author replies in kind:

> His reference to the ". . . present state of the art . . ." of electron microscropy is intriguing. Glutaraldehyde-OsO4 TEM [transmission electron microscopy] has been around since before I was in junior high school, and I have never considered it an especially difficult "art." Maybe I have just been blessed with cooperative tissues that prefer to die gracefully? (The c___ tissues, I mean, my own personal tissues will go down fighting!)[2]

Although present in author and referee discourse, emotion is entirely absent from author-editor correspondence concerning these papers, papers accepted contingent upon revision. First of all, correspondence of this sort represents direct contact with strangers: politeness is the general rule. But if authorial politeness is clearly deferential, editorial politeness is more complex in its motivation. It is doubly involuted: accompanying referee reports, it is a demand disguised as a request, and a demand from the editor disguised as a demand from the referees. Note the difference in tone between an editorial letter to the author and its accompanying folder-note, meant only for private consumption:

> We are enclosing the comments of two reviewers and one copy of your manuscript. Both reviewers offer many suggestions, which should be carefully considered and incorporated into the revision.

> While both reviewers recommmend publication, both reviewers have made so many important and detailed comments and suggestions that only a revised manuscript can be accepted for editing. [signature of one of the editors]

The polite fiction in the letter anticipates and mitigates any authorial wrath and deflects any remaining negative feelings away from the visible editors and toward the anonymous referees.

There are two additional criteria of the ideal speech situation: all interlocutors must be free to use any speech act, and they must have equal power over the exchange. But it is clear that authors are prohibited from issuing commands and inhibited from asking critical questions, while editors and referees do both freely. This difference is an index of the power of the latter and the relative powerlessness of the former: because they are making a request, authors know that "[it] is not obvious that [editors] will [publish a paper] in the normal course of events of [their] own accord" (Searle 1969, p. 66). Moreover, power is divided unequally in less obvious ways. In physics, for example, referees as a group have a higher professional status than do authors; in Zuckerman and Merton's study "12 percent of the referees [those of higher rank] contributed one-third of all referee judgements"; in addition, their judgments were somewhat harsher than those of their lesser-ranked colleagues (Zuckerman and Merton 1973, pp. 483, 490). These facts are consistent with the notion that referees and authors need not be equally constrained by the prospects of role reversal.

Furthermore, the general rule of referee anonymity forces authors to frame a reply in ignorance of their interlocutors. Of course, to balance matters, authors may also be anonymous, but genuine anonymity for authors may well be an ineffectual charade: any worthwhile research program is an intellectual fingerprint. Indeed, the most thorough study of scientific peer review, that of the National Science Foundation, concludes that "'blinding' of proposals is both impractical and ill-advised."[3] Finally, since multiple submission is seen as unethical, authors are forced to endure delays beyond their control, delays which may cause them, but not their editors or referees, considerable anxiety, and which may easily reduce the value of their submission as a contribution to ongoing scientific debate.

In addition to these criteria, the ideal speech situation "must permit a progressive radicalization of the argument; there must be the freedom to move from a given level of discourse to increasingly reflected levels" (McCarthy 1982a, p. 305). As general rule, however, peer review blocks discussions of its own fairness and of the ultimate value of the knowledge to which it gives preliminary certification. For example, editorial decisions are without formal appeal. In fact, any appeal is unusual and must be motivated by exceptional circumstances, such as the allegation of systematic referee bias. In appeal, of course, it is not the decision that is at issue; it is the process by which the decision was reached (Habermas 1979, p. 64; Pinch 1985, p. 182).

Finally, the lack of reliability among referees' and editors' judgments counts against the alleged rationality of consensus. Though we may concur that global judgments by referees are reliable in the sense that "two scientists acting unknown to each other as referees for the publication of one paper usually agree about its approximate value" (Polanyi 1964, p. 51), we may still doubt whether referees would agree concerning the basis of their agreement: in my admittedly small sample, there was consistently little overlap between the questions and comments of two referees of a single paper.

The Correction of Deviation

The criteria for the ideal speech situation have equality of power as their common denominator. Although peer review in science clearly deviates from this principle, it also attempts to correct for this deviation. In the pursuit of a balance of power between authors and referees, editors generally assume that the rejection of a paper depends on a clear negative decision on the part of both referees; a split decision ordinarily favors the authors.[4] In conveying the results of peer review, moreover, editors take care to shield authors from the worst of referees' excesses. For example, one referee begins his report by labeling the paper's Abstract "incomprehensible," and ends with the following peroration:

> The entire paper appears to be a fantasy. The methods used . . . are not at all conducive to understanding the developmental anatomy of this s____. In my opinion, the paper is unscientific and is not worth publishing. The paper makes no contribution to the solution of the problem raised in the last paragraph of page 2 or any other problem.

Because they were permitted to savor this report uncensored, the authors could fully appreciate the editors' inventive tact: "Both reviewers," the editors aver, "offer constructive criticism, and [the other reviewer's] suggestions appear especially helpful."

In the case of another paper, the criticisms of the referee seem devastating:

> The authors contradict themselves on a very important point. . . . a more rigorous and thorough analysis of the methodology for enzyme cytochemistry [chemistry of cells] is in order. Since this has not been performed prior to submission of the present manuscript, it is essential that it be done before the manuscript should be accepted for publication. As the manuscript stands, it is not much more than a compilation of a few observations that have limited and questionable application to biological problems. At best, the data represent interesting observations with little significance and limited importance.

The authors, no doubt, would have been upset had they experienced this merciless assault; however, in this case, they received only one, essentially favorable, referee report.

Whatever the motives for these hostile reports—and in one case at least, on the basis of evidence not presented here, conflict of interest may be adduced—they plainly reflect not on the papers as responsible science, but on the referees as responsible members of a scientific community. Like referees who must be badgered to return reports in a timely mannner, those who misuse their temporarily bestowed power are likely to be stricken from editorial rosters.

In the case of papers accepted contingent on revision, the referees, originally retained as judges, are redefined by editors as repairers and collaborators; authors, though still defendants, have a new, editorially defined task: the reestablishment of their authority against referee requests for alteration and amplification.

This change in relationship does not mean that authors must comply fully with referee requests. Certainly authors must consider these carefully, but the admitted failure to provide satisfactory answers to referee questions is no bar to publication. For example, a referee asks: "Is cell death involved in the formation of some portions of the hollow s____?" In their reply, the authors confess their failure to find a satisfactory answer: "We have gone through our slides once again to find out if death is involved in the formation of some portion of the hollow s____ . . . . However, we could not get enough information to make a categorical comment on this aspect. We have, therefore, not included [the] cell degeneration aspect in the manuscript."

Indeed, publication can proceed even when there is a serious disagreement with a referee concerning the fundamental assumptions of the field. For instance, a referee provides support for the contention that "the systematics section [of this paper] . . . needs some rethinking":

> For historical reasons and due to the thorough studies on Ar____ and Rei____ by the author he would like to erect the order Ir____ in which it is proposed to include Ib____. Ib____, however, [is] known as a whole _____ including external h____, sp____, etc.

The author responds by roundly attacking this last assumption:

> Although I do admit that Ib____ is better known than the other genera in terms of overall morphology, it is very far from being completely understood as a whole _____. S____ and B____ (1973, p. 368) were careful to point out the limitations of their evidence on external morphology especially in relation to the poorly preserved structures thought to be sp____. Essentially nothing is known about the histology of these structures, and on the basis of the evidence presented in their paper, even the occurrence of sp____ can hardly be accepted as irrefutable fact. Information on the anatomy of Ib____ is even more fragmentary. In fact, I wouldn't at all be surprised that when more is known internally about Ib____, it will be shown to be synonymous with Ar____ which at present is far better known internally.

In authors' responses, one sees two central strategies of defense. In the excerpts just quoted, authors disagree by plausible counterargument, citing, as do lawyers, whatever authorities suit their purpose, including the authority of reobservation. In a second strategy, authors side, wherever possible, with the authority of the editors or of the positive referee. Concerning a referee's request for additional electron micrographs, an author says: "Nothing infuriates me more than published unnecessary TEMs [transmission electron micrographs] . . . If you, as editors, really want them, I will supply them in quantity, but otherwise I think they are a waste of space." Another author summarizes as follows: "I can only conclude that Reviewer A objects either to my inclusion of taxa on the basis of unique and derived features, or to my use of a new name, or both. To be perfectly honest, I am greatly surprised by the difference in the response of reviewers A and B on precisely these points. My approach is understandable in the paper, I believe, and reasonable."

But dissent is far from the rule, since dissent risks rejection. Ordinarily, authors reestablish their authority by treating referees as collaborators—as additional, if invisible, authors of the paper; indeed, sometimes they acknowledge in print the help of these still anonymous critics. Authors also comply patiently with requests as time-consuming as an omnibus reexamination of slides; they back down gracefully in the face of error, and willingly curb unwarranted speculations to the extent, in one case, of creating a new major division: "A Limitations and Recommendations section was added," an author asserts, "to explain the limitations of our study and summarize the findings."

Editors are themselves limited in power by the professional situation in which they find themselves. All wish to publish journals that will be widely read and frequently cited. Though they do not rely on any one author, they do depend on authors generally to supply a steady stream of acceptable manuscripts. Equally, authors do not rely on one editor's decision; they have the power of the free market for ideas in their field of study. As long as there exist other journals of approximately equal prestige to which rejected authors may send their papers, editors are faced with the prospect of seeing in print, or worse, in the Science Citation Index, papers they judged unsalvageable for their journal. None of these considerations is decisive in any particular case; however, in concert, such considerations set serious limits on the irresponsible use of editorial power.

Necessary Deviations

Peer review in science is a system of communication that achieves rational consensus by means of the balance it strikes between deviations from the ideal speech situation and their correction. Since these deviations prevent the full realization of the ideal speech situation, it might seem that fully rational consensus depends on their total elimination: in an ideal society, scientific peer review would permit unimpeded authorial initiative, endless rounds of give and take, unchecked openness among authors, editors, and referees. Until these circumstances prevailed, peer review communication could not fully realize rational consensus. But peer review is an institutionalization of the ideal speech situation, a practical matter: the question is not whether it embodies a full realization of the ideal speech situation, but whether it "can be justified as leading to outcomes that would, ideally, be agreed to in unconstrained dialogue."[5]

This is not to deny the initial attractiveness of the full realization of the ideal speech situation in peer review, a realization that would ensure both the quality of accepted papers and the general intellectual and political health of the community. At the same time, this realization would guarantee that all potentially worthwhile research results would be shared with appropriate peers, that proper credit would be given, and that a scarce resource—journal space—would be responsibly allocated. Unfortunately, placed back in its proper context, the scientific form of life of which it is a part, the pursuit of fully rational consensus competes with an equally important good, an equally scarce resource—professional time.

The need to conserve professional time is an imperative in every discipline: it explains the strong constraint against multiple journal submissions of the same paper; it also accounts for the powerful resistance to otherwise sensible suggestions for peer review improvement—for example, the suggestion that referee reports be shared with authors *before* the editorial rejection of a paper. A decided movement in the direction of the ideal speech situation would improve an author's position only at the expense of the professional time of editors and referees. By substituting review for research, self-reflection for science, such a movement would also impede the creation, and the timely communication, of potentially valuable science.

The full realization of the ideal speech situation is necessarily limited. This limit stems from a "less or more uneasy compromise between principles which in their extreme form cannot coexist" (Berlin 1978, p. 102). The ideal speech situation cannot be viewed as a goal toward which we gradually move because such a construal involves us in paradox. We must suppose that rational consensus is a good we can pursue without regard to its possibly undesirable effects on other, equally desirable goods. In other words, we must subscribe to the view "that all good things [are] certainly compatible, and indeed interlocked, with each other" (Berlin 1978, p. 95): we must subscribe to an Enlightenment myth. For Habermas, however, particular social institutions like peer review are practical matters: "It is a question of finding arrangements which can ground the presumption that the basic institutions of the society and the basic political decisions would meet with the unforced agreement of all those involved, if they could participate, as free and equal, in discursive will-formation" (1979, p. 186).

Thus the heedless pursuit of rational consensus would have an unavoidable, and undesirable, cost: the ideal speech situation cannot be fully realized without devaluing professional time and progressively undermining peer review's central purpose, the timely certification of new science. For this reason, every time a new scientific or scholarly journal is founded, peer review is consensually reinvented with virtually identical constraints, and with virtually the same balance between deviations from the ideal speech situation and their correction.

Publication and the Status of Scientific Knowledge

The rational consensus that peer review exemplifies has epistemological implications. Scientific reports and peer review documents are very different illocutionary enterprises. Scientific reports are cognitive; they thematize truth, the link between statements and the world; their dominant speech act is the confirmative, an expression by which the writer indicates that his beliefs come about "as a result of some truth-seeking procedure, such as observation, investigation, or argument."[6] In both experimental and theoretical papers, these truth-seeking procedures are designed to bestow a sense of certainty upon conclusions. In contrast, peer review documents are regulative; they thematize trust, the intersubjective nexus linking authors, editors, and referees. In peer review, then, the results of a cognitive process are certified, at least initially, by a regulative one. This certification necessarily diminishes the epistemological status of the science it certifies: cognitive claims are judged not by testing against the world, but procedurally, by peer consensus. The knowledge that peer review certifies is grounded wholly in argument: "Post-empiricist philosophy of science has provided good reasons for holding that the unsettled ground of rationally motivated agreement among participants in argumentation is our only foundation—in questions of physics no less than in those of morality" (Habermas 1982, p. 238).

After peer review, claims that have been vigorously disputed are restored to their initial status; they are turned again toward the world as meant by science; the machinery of peer review disappears from sight. Publication performs this task; it frees the scientific report from the epistemological liabilities of peer review. Publication is the symbolic act that obliterates all traces of the procedure by which the knowledge it asserts is certified; by this obliteration, publication renews, at least temporarily, the credibility undermined by peer review: claims are again wholly about the world. Once they have served their purpose, peer review documents become generally unavailable to scrutiny; scientific reports are read only in the context of other reports—the official literature of science.

As a result of this process, there is cultivated a systematic neglect of the relationship between the claims in these reports and the process by which their truthfulness is initially certified: the move to publication systematically distorts the wholly argumentative grounding of the knowledge that peer review certifies. In the public domain, all that is visible is the scientific report, another step in the steady march toward certain knowledge.[7] In Habermas's terms, this is an institutionalization of systematically distorted communication: "In such cases at least one of the parties is deceiving himself about the fact that he is acting with an attitude oriented to success and is only keeping up the appearance of communicative action" (1984, p. 332).

Rival Explanations

The ideal speech situation provides one explanation of peer review; functionalism and the strong program in the sociology of science provide two others. Unlike their more conservative brethren, the strong programmers refuse to confine their analyses to the social structure of science, the traditional subject matter. Instead, they train their sights on the knowledge claims that scientists make; they insist that these claims are also subject to sociological explanation. This research focus need not mean the abandonment of the claim that science is a rational enterprise, that rational consensus is the *telos* of both peer review and science; still, in some more radical models of social constructivism—in Woolgar and Latour's *Laboratory Life* and in Latour's *Science in Action*, for instance—rationality is not so much explained as explained away. For Latour and Woolgar, the explanatory trope is economic power: "The set of statements considered too costly to modify constitute what is referred to as reality," a reality in whose creation rationality has had no part (1979, p. 243).

For Latour, a decade later, the trope alters to physical power: science is a means of "act[ing] at a distance on unfamiliar events, places, and people." If we can render these events, places, and people mobile, stable, and combinable, if "they can be cumulated, aggregated, or shuffled like a pack of cards . . . then a small provincial town, or an obscure laboratory, or a puny little company in a garage, that were at first as weak as any other place will become centres dominating at a distance many other places" (1987, p. 223). Such explanations abandon the difference between rationality and irrationality: "All this business about rationality and irrationality is the result of an attack by someone on associations that stand in the way" (Latour 1987, p. 205); "if we wish to continue the study of the networks of technoscience, we must straighten up the distorted beliefs and do away with this opposition between rational and irrational ideas" (p. 185).

Peer review may be analyzed according to these imperatives. In a paper in this vein, Myers asserts that peer review negotiates the level of claims permissible in a scientific paper. The higher the level, the higher the paper's status; the higher the status, the more difficult the negotiation. To support his view, Myers recounts the publication histories of two biological papers. In each case the author tries to obtain acceptance for the broadest possible claims in the most prestigious professional forums: "Both the biologists I am studying try to make the highest level claim the editors and reviewers will allow" (Myers 1985, p. 602).

According to Habermas, however, we err when we reduce communicative action to a contest for power: "Generalized forms of communication such as influence and value commitment require illocutionary acts and thus remain dependent on the binding effects of using language with an orientation to mutual understanding. [In contrast,] steering media such as money and power guide interaction through ego's intervention in the situation of alter, through perlocutionary effects [the results of speech acts oriented only to success] if need be" (1987, p. 280; see also 1984, p. 292). But to equate communicative action with power is to disallow a crucial distinction, the "strict distinction between *ties that are motivated empirically* through inducement and deterrence and *trust that is motivated rationally* through agreement based on reasons" (p. 280; his emphasis). To equate communicative action with power is to remain insensitive to real differences between rationality and irrationality, between communicative action oriented toward understanding and systematically distorted communication. Analyzed in terms of power only, peer review is, quite simply, unrecognizable.

In a dismissive aside, Myers essays a third explanation of peer review data, a functional explanation that would "see in the relegation of [these] articles to more specialized journals an example of how the publication process works, protecting these researchers in other fields from just this kind of claim from outside their own research programmes, and thereby preventing the capricious redirection of goals, the proliferation of research programmes and the scattering of resources" (1985, p. 623). But to radical social constructivists like Latour and Woolgar, such explanations explain nothing. Only power is real: functions and rationality are equally epiphenomenal. From a Habermasian point of view, functionalism is defective for a different reason: its self-regulating systems can account for communicative action only by the reduction of such action to a systems function with which it differs in essence.

Unlike the ideal speech situation, neither functionalism nor radical constructivism explains rationality: the latter dismisses it as an artifact, the former equates it with system self-maintenance.

Debates about rationality follow a common pattern. Rationality is described in terms of its alleged necessary conditions—arguers must be consistent, must heed the law of contradiction, must be constrained by *modus ponens*. Relativists then attack these purported universals with supposed counterinstances (for example, Hollis and Lukes 1982). The real need, I think, lies elsewhere: we require a definition of rationality that is not vacuous, one that at the same time takes into account the presence of "divergent rationalities": the intuition that the divergent fundamental beliefs and patterns of inference of other cultures do not count as irrationality but as rationality of another sort (Shweder 1986).

The ideal speech situation gives us such a definition. Because this theoretical construct defines rationality as a process, not a product, it insists on convergence only as a regulative ideal; in practice it expects diversity, and is embarrassed neither by divergent rationalities among distinct cultures (Shweder 1986, p. 188–190) nor by seemingly wide differences within particular cultures, differences such as the diverse referee judgments that generally characterize peer review. Thus the ideal speech situation permits a rhetorical definition of peer review in which rationality still has a central place.

CHAPTER 10

The Origin of *The Origin*

On October 2, 1836, after its circumnavigation of fifty-five months, the *Beagle* anchored off the English coast. In Darwin's kit was a small book bound in red leather, the *Red Notebook,* two-thirds full of geological notes. Busy disposing of the specimens he had gathered on the voyage and reacquainting himself with his family, Darwin did not immediately focus his attention on writing and thinking. But by the beginning of the following year, he was simultaneously converting his diary of the voyage into a travel book and making additional entries in the *Red Notebook.* It is these entries, these "brilliant scraps" (Herbert 1974, p. 247) made in the first half of 1837, that are the subject of the rhetorical analysis of this chapter. They contain Darwin's first speculations on evolutionary theory and lead directly to his decision to begin the first of his transmutation notebooks. To read these notes through is to experience the birth of a great idea.[1]

Self-Persuasion as Rhetoric

Rhetorical analysis may be appropriate for public texts meant to persuade particular audiences; but what warrant can there be for applying the classical categories of style, arrangement, and invention to the *Red Notebook*—these jottings, meant apparently for no audience, haphazardly arranged, and so abbreviated as to be, in many cases, barely comprehensible? The warrant rests on a theory of self: "that the self owes its form and perhaps its very existence to the circumambient social order" (Harré 1984, p. 256). The first step in its creation is the formation of the person, the individual public being derived by means of "psychological symbiosis" from the social order.

In this process, "one person [supplements another's] public display in order to satisfy the criteria of personhood with respect to psychological competencies and attributes in day-to-day use in a particular society" (p. 105).

It is from the person each of us becomes that the singular inner being we call our self emerges: "Our personal being is created by our coming to believe a theory of self based on our society's working concept of a person" (Harré 1984, p. 26). If the self is a theory we come to believe, its essence is rhetorical; *a fortiori* particular networks of belief (such as scientific theories) must be the products of rhetorical transactions in the self's arena.

This view of the self is also in accord with our everyday metaphors of mentation. In the latter half of the *Red Notebook*, Darwin, we might say, was of two minds concerning evolution; in a series of entries, he debated with himself over its possibility. Charles Sanders Peirce turns these metaphors into philosophy when he says: "A person is not absolutely an individual. His thoughts are what he is 'saying to himself,' that is, is saying to that other self that is just coming into life in the flow of time. When one reasons, it is that critical self that one is trying to persuade . . . [moreover] man's circle of society (however widely or narrowly this phrase may be understood), is a sort of loosely compacted person, in some respects of higher rank than the person of an individual organism" (1955, p. 258).[2]

Biographers, historians, philosophers of science, cognitive psychologists—all have exhibited a legitimate interest in Darwin's Notebooks as evidence of his intellectual development. But none has consistently treated these entries as a special genre: not public documents or letters meant for identifiable audiences; not diaries framed for the particular audience of one's future self; not first drafts of such documents, early attempts at audience accommodation; not Rorschach or TAT protocols, self-presentations to an audience of psychologists. The audience of the Notebooks is Darwin, Darwin immediately, and Darwin only; their assertions are speech-acts of a special sort, acts whose epistemological status is subject to marked fluctuation. I contend here that the fluctuating epistemological status of the speech-acts in the *Red Notebook* is an essential aspect of their correct interpretation, an aspect best captured by the rhetorical analysis that follows.

Style as an Index of Self-Persuasion

There are two styles in the *Red Notebook*: the cryptic and the telegraphic. Although the telegraphic style maintains most of the norms
of Victorian prose, it frequently omits articles, copulas, and the exis-
tential "there"; and it regularly employs standard abbreviations:

> Volcanos only *burst* out where strata in act of dislocation (NB. dislocation
> connected with fluidity of rock ∴ {in earliest stage} when covered up
> beneath ocean).—The first dislocations & eruptions can only happen
> during first movements, and therefore beneath ocean, for subsequently
> there is a coating of solidifying igneous rocks which would be too thick
> to be penetrated by the repeted trifeling injections.— (1987a, p. 68)[3]

In addition to the telegraphic features, the cryptic style displays an
absence of connectives, especially logical connectives; deviations from
syntax radical enough to render semantic connections problematic;
and radically deviant punctuation:

> Speculate on neutral ground of 2. ostriches; bigger one encroaches on
> smaller.— change not progressif<e>: produced at one blow. if one
> species altered: <altered> Mem: my idea of Volc: islands. elevated.
> then peculiar plants created. if for such mere points; then any mountain.
> one is falsely less surprised at new creation for large.—Australia's = if
> for volc. isl$^{d.}$ then for any spot of land. = Yet new creation affected by
> Halo of neighbouring continent: ⵌ as if any |
>
> creation {taking place} over certain area must have peculiar character:
> (1987a, p. 61)

The telegraphic style is a shorthand, designed like all shorthands
to close the gap between the rate of verbal production and its tran-
scription. But the cryptic style has more complex roots. Built up
chunk by linguistic chunk, each framed by terminal or near-terminal
punctuation, passages in this style actively resist interpretation.[4] It is
difficult to see how such linguistic strings could have issued from the
operation of syntactic rules on semantic components stored simultaneously in short-term memory. More plausibly, they represent a stage
prior to the application of these rules. Under this construal, each
chunk must have been retrieved individually with effort, must have
been simultaneously *thought* and *uttered*.

Consider these two utterances separated by less than a hundred words:

Mem[ento]. SUBSIDENCE Uspallata of which no trace except by trees

Let it not be overlooked that except by trees, I could not see trace of Subsidence at Uspallata.—[5]

According to my reading, on the self-command—"Remember"—the four linguistic chunks in the first utterance are called forth sequentially from long-term storage, the second by the first, the third by the second, the fourth by the third. On this interpretation, to read the *Red Notebook* is to experience the psychological reality of a generative grammar: the second linguistic string, the sentence, derives from the first by means of psychological rules analogous to the syntactic rules of Chomsky and his followers.

This view of the cryptic style is consistent with current theories of mental processing. It does not depend on any theory of what thought is, but finds plausible Kintsch's notion that thoughts are stored in long-term memory in propositional form (Klatsky 1980, pp. 196–199), and Fodor's speculation that there must be a system of internal representations, an internal code, at least as complex as natural language—complex enough, certainly, to "represent representations" (1975, p. 172; see also p. 156). If Fodor is right, the cryptic style may be evidence for the psychological reality of such a code. Other interpretations of this style are certainly possible; but all must take into account the fact that Darwin's words, while conceivably in the order of thought, deviate markedly from the norms of sequence of ordinary speech.

In the *Red Notebook* we reach an important limit beneath which rhetorical analysis cannot operate. Entries in the cryptic style approach a threshold below which interpretation must give way to causal analysis: language much more disordered than this would cease to fulfill the norm of comprehensibility that is one of communication's necessary conditions. A style more disordered than the cryptic would also be excluded from rhetorical theory because "someone cannot have a belief unless he understands the possibility of being mistaken, and this requires grasping the contrast between truth and error—true belief and false belief. But this contrast . . . can emerge only in the context of interpretation, which alone forces us to the idea of an objective, public truth" (Davidson 1984, p. 170; see also Quine 1976, p. 235). Davidson does not equate thought with sotto voce speech; the claim is that complex belief is possible only to creatures who have speech. Any other view must rely on notions of belief so broad as to be nearly vacuous (for example, Dogs *believe* there is a fox in that bush).

In the *Red Notebook*, as Darwin's speculations approach the stability of his final formulations, his style approximates the norms of his published prose. As they approach these norms, his utterances simultaneously approach the norms that speech-act theory stipulates for assertions in the public domain: that they be clear, that they be true, that we take responsibility for their truth.[6] Thus the epistemic status of entries in the *Red Notebook* varies: the nearer these speech-acts approach the cryptic style, the closer they approximate the suspension of the validity claims common to public assertions. This suspension indicates that, early in the development of his ideas, Darwin is prepared to fix neither reference nor belief. Neither confusion nor irresponsibility is at work; on the contrary, only so long as he remains uncommitted to the full truth of his assertions can Darwin maintain the prolonged suspension of belief essential to the acceleration of internal conceptual change.[7] In Darwin's private mental space, style abandons its public function; it is no longer a means of persuasion. Because it provides evidence for the epistemological status of entries, style becomes, instead, an index of the relative stability of the beliefs Darwin espouses.

Arrangement as an Anticipation of Intellectual Development

Arrangement in the *Red Notebook* is also an index of the progress of Darwin's thought: arrangement anticipates his intellectual development. Themes later linked in important theoretical formulations appear separately, form sequences, and become intertwined.

In the entries in the *Red Notebook* made soon after his return, Darwin focuses on a range of topics that were to form two of the three *leitmotifs* of his intellectual career: the causes of geological change and the nature and origin of species:[8]

I. The geological and mineralogical character of South America and Pacific (as compared to Europe)
    A. Igneous change
        1. Subsidence and elevation of the earth's crust
            a. the creation of mountains, volcanoes, and earthquakes
            b. their pattern
            c. the action of volcanoes and earthquakes
        2. the action of lava
    B. Aqueous change: the action of the ocean
        1. silting to build up land
        2. denudation to wear down land
        3. transportal of boulders
        4. glacial action
II. Biogeography of South America and Pacific
    A. Distribution of flora and fauna in space
        1. effect of climate
        2. effect of barriers
        3. question of adaptation
    B. Distribution of flora and fauna in time
        1. extinction of species (fossil record)
        2. creation of species: descent

Although actual entries are far less organized than this outline suggests, their order exhibits trends that anticipate Darwin's mature formulations. Here is a representative sequence of thirty-one entries, coded according to the outline and arranged into fourteen thematic groups:

IA, IA,
IIA,
IA, IA,
IIA,
IIB, IIB, IIB, IIB,
IIA,
IIB,
IIA, IIA, IIA, IIA, IIA, IIA,
IIB, IIB,
IA,
IIB, IIB,
IIA, IIA,

IB,
IA, IA, IA, IA, IA[9]

In this display two contrasting tendencies are apparent. According to the first, Darwin intersperses the geological and the biogeographic; according to the second, he pursues the same theme in a single run of entries. The tendency to intersperse anticipates the close connection between geology and biogeography in Darwin's mature theory; the tendency to separate anticipates Darwin's decision to bifurcate his intellectual efforts: after filling the *Red Notebook,* he simultaneously began Notebook A on geology and Notebook B on transmutation.

There is a third tendency that no outline can represent: short, isolated entries on particular themes are followed by longer passages in which these themes appear in concert. Witness the first two biogeographic entries in the run IIA, both on the distribution of flora and fauna in space:

> 1. Webster Antarctic veg:—
> 2. no. mad dogs. Azores. although kept in numbers.
> (1987a, p. 60)

These are followed by a longer biogeographic entry concerning fossil deposition: the distribution of flora and fauna in time. In this, Darwin speculates that a worldwide drought may have led to the common burial of species with normally divergent territories:

> Consult W. Parish. & Azara about dry season[.] 1791. seen commonly bad over whole world. {Was it so in Sydney, consult history? Phillips.} | 1826.27.28 grt. drought at Sydney. which caused Capt. Sturt expedition.—| ¿ Another one in 1816(?).—[10]

Four pages later Darwin links distribution through space and time in a powerful single formulation:

> The same kind of relation that common ostrich bears to (Petisse. & diff kinds of Fourmillier): extinct Guanaco to recent: in former case position, in latter time. (or changes consequent on lapse) being the relation. As in first cases distinct species inosculate [separate], so must we believe ancient ones: {∴} not *gradual* change or degeneration. from circumstances: if one species does change into another it must be per saltum [by leaps]—or species may perish. = This <inosculation> {representation} of species important, each its own limit & represented.—Chiloe

creeper: Furnarius. <Caracara> Calandria; inosculation alone shows not gradation:—(1987a, pp. 62–63)

In this passage we have moved past thematic connections into a relationship of natural necessity, a theory about the natural world. This traversal from theme to theory is tellingly exemplified by the $\therefore$ symbol that Darwin added as an afterthought, a "therefore" that marks the sudden inception of his awareness of a necessary relationship, a natural law.

As the arrangement of passages where themes intertwine approaches the stylistic and organizational norms of Darwin's finished prose, their theoretical content verges on the norms of his mature theory: the distribution of flora and fauna in space and time subject to the same natural law. When we examine the arrangement of the *Red Notebook*, then, we see themes become theory. But in the passages in which themes previously separate are closely linked, we leave the province of arrangement proper and pass over into invention.

The Play of the Intellect

From Theory to Theory. In his *Autobiography*, when Darwin reflects on his most important intellectual achievement, evolutionary theory, he means by "theory" not a disjointed collection of thematically associated lower-level principles but a fully articulated set of beliefs about the origin of species, one that explains all of the facts within its compass, including those that may seem anomalous at first glance. He means a theory he can endorse publicly, one to which he can fully commit his scientific reputation. Only at the end of 1838 did he feel he had such a theory in hand.[11]

In its pursuit Darwin's thought passes through three stages, each dominated by differing heuristic imperatives. During the voyage, Darwin collects specimens. Although from time to time he is aware that their contemplation may lead to novel conclusions, he has developed no theory of their origin. From his return until 1840, he passes through a second stage: in the Notebooks he ponders his collections and speculates continually about their implications for theory. This stage reaches its climax with the Malthusian revelation: by its means, Darwin moves from the belief "that species [are] mutable productions" to an understanding of a *vera causa* for this phenomenon; at

last, he has "a theory by which to work." Before Malthus, he had
"worked on true Baconian principles, and without any [comprehen-
sive evolutionary] theory collected facts on a wholesale scale" (1958,
pp. 119, 120, 130). Beginning in 1839, he turns to the final stage of
his work: the full articulation of his views and their public
presentation.[12]

This view concerning the second, most creative stage of Darwin's
intellectual development deviates markedly from the current consensus that the Notebooks contain a sequence of predecessor theories,
complete evolutionary explanations that Darwin devised and discarded as he moved toward his final formulation.[13] Gruber, for
example, claims that Darwin at one time formulated a monad theory:

> If the species are being constantly created, they must constantly be
> destroyed: "this tendency to change . . . requires deaths of species to
> keep numbers of forms equable" [1987a, pp. 175–176]. Elementary
> forms of life, "monads are constantly formed . . ." [p. 175] and these
> have a definite life span. While a monad lives, it undergoes evolution:
> "the simplest cannot help becoming more complicated. . . ." [p. 175].
> When the monad's life is over, it dies, and with it must die all the species
> that it has become [p. 177]. (1981, p. 136)

In his search for a Darwinian monad theory, Gruber represents
private meanings as public assertions. But in the Notebooks, it is the
exploratory nature of these assertions that is emphasized:

1. Species balance is no sooner mentioned than it is questioned:
   **"but is there any reason for supposing number of forms
   equable**[?]**."**[14]
2. The clause "monads are constantly formed" is the antecedent
   of a hypothetical.
3. The possibility that monads have a definite lifespan is considered only to be rejected almost immediately.
4. The thought behind Gruber's last sentence is embedded in an
   utterance that wholly denies its truth.[15]

Kohn founds his case for the existence of predecessor theories on
the following definition: "Theories can be construed as attempts to
capture the essentials of the subject area which they explain."[16] With
this definition as his guide, he shows that entries in the Notebooks
can be reconstructed to form networks of belief about the origin of

species. It is these networks that Kohn identifies with predecessor theories. Although Kohn's reconstructions are indeed theories, they are not Darwin's theories: they identify as theories formulations that Darwin regarded as entirely tentative.

To seek and exhibit predecessor theories in the manner of Gruber and Kohn misconstrues the character of the Notebooks. Those who claim the existence of predecessor theories commit Darwin to ideas that he expressed in an essentially private language virtually free from commitment, a language appropriate to the intellectual freedom that is a precondition of creative endeavor. "I dwell in Possibility – / A fairer House than Prose –" writes Emily Dickinson (1963, II, 506). In the Notebooks, it is in this house that Darwin dwells: themes are set in rapid mental motion, theories are tried on for size. This spirit of intellectual play dominates the most creative stage of Darwin's development.

During that stage, Lyell's great work, *Principles of Geology,* swept Darwin off his intellectual feet: "I am become," he wrote to William Darwin Fox in 1835, "a zealous disciple of Mr Lyell's views" (1985–, I, 460). Indeed, theory after theory in the *Red Notebook* was not devised by Darwin but borrowed from Lyell—the common cause of volcanoes and earthquakes, the geological impact of water in motion, the effect of surrounding ocean in mitigating island and coastal climates. Most important, Darwin held, as did Lyell, that the causes of past geological change must operate in the present at the same general level of intensity.

But Darwin's discipleship was far from passive; it did not prevent him from challenging his master. "I am tempted," he says in the same letter to Fox, "to carry parts [of Lyell's theories] to a greater extent, even than he does." In print Darwin would challenge Lyell's theory of the formation of coral reefs; in private he would question his theory of the origin of species, a topic integral to Lyellian geology.

In the second book of the *Principles,* Lyell considered and rejected evolution:

> If a tract of salt-water becomes fresh by passing through every intermediate degree of brackishness, still the marine molluscs will never be permitted to be gradually metamorphosed into fluviatile species; because long before any such transformation can take place by slow and insensible degrees, other tribes, which delight in brackish or fresh-water,

will avail themselves of the change in the fluid, and will, each in their turn, monopolize the space. (II, 174)

In the *Red Notebook* Darwin puts into intellectual play the evolutionary premise that Lyell so decisively rejects: "As in first cases distinct species inosculate [separate], so must we believe ancient ones: {∴} not *gradual* change or degeneration. from circumstances: if one species does change into another it must be per saltum." Why *must*? If Lyell's argument against gradualism is sound and general (and not the petitio principii it turns out to be, assuming what it set out to prove), evolution must occur by leaps: evolution entails saltation. In this instance Darwin places Lyell's theory into intellectual play; in so doing, he derives a theoretical formulation of his own. Not a theory: Darwin is committed to a prolonged exploration that prohibits early closure.

From Fact to Theory. Around August 1833, still in the stage of collection, Darwin found himself in Northern Patagonia. While there he heard of a rare form of ostrich,[17] the Avestruz Petise, a bird smaller than the common ostrich, darker in color, shorter of leg. Around Christmas of the same year he sat down to a dinner of an ostrich shot by Conrad Martens, the ship's artist. At first Darwin thought the bird was a young specimen of the common sort; after the meal, however, his memory of the Petise returned, and he saved the head, legs, and several feathers of the bird he and his shipmates had just consumed. In notes probably written within three months of the incident, Darwin recounts his change of mind. He points out the differing range of the smaller bird, and states: "Whatever Naturalists may say, I shall be convinced from such testimony. as Indians & Gauchos, that there are two species of Rhea in S. America" (1963, p. 274).

At the time, Darwin collected the specimen as part of his routine duties as the expedition's naturalist. By March of the following year, however, he had singled out the species for special consideration: "But what is of more general interest is the unquestionable (as it appears to me) existence of another species of ostrich, besides the Struthio Rhea.—" (1985–, I, 370; see also 1977, p. 8). Two years later, in mid-1836, on the last leg of Darwin's journey home, the Petise appears again in a series of ornithological notes: "In conclusion, I may repeat that the Struthio Rhea inhabits the country of La

Plata as far as a little south of the R. Negro, in Lat. 41°: & that the Petise takes its place in Southern Patagonia, the part about the R. Negro being neutral territory" (1963, p. 272; see also 1987b, p. 109, and Sulloway 1982, p. 337).

On March 12 of 1837, in his first year in England, Darwin reported that the ornithologist John Gould had confirmed his identification, naming the new species in honor of its discoverer (1985–, II, 11). Two days later Gould reported his identification to the Zoological Society; at the same meeting, Darwin read a paper on the subject. He avoided speculation, stating, according to the Secretary, "that the *Rhea Americana* inhabits the country of La Plata as far as a little south of the Rio Negro, in lat. 41°, and that the *Petise* takes its place in Southern Patagonia" (1977, p. 40). At this point the newly named *Rhea darwinii* disappears from Darwin's public writing.

But not from his private writing. In the days following his presentation to the Zoological Society, Darwin initiated in private the evolutionary speculation he eschewed in public: he recorded in the *Red Notebook* an entry singling out the neutral ground between the ranges of the two species as a site for theorizing (Sulloway 1982, p. 381): "Speculate on neutral ground of 2. ostriches." It was only a fact that two species of ostrich shared contiguous territory; but it was a fact destined to have a profound impact on Darwin's thought. In his *Autobiography,* he says of his most creative period: "It was evident that such facts as these . . . could be explained on the supposition that species gradually become modified; and the subject haunted me" (1958, pp. 118–119).

Within the next two months, the first product of this speculation comes into view. It concerns what is to become a recurring theme, descent with modification: "When we see Avestruz two [ostrich] species. certainly different . . . Yet one is urged to look to common parent? why should two of the most closely allied species occur in same country?" A slightly later entry contains a far more assertive reference to descent with modification: "I look at two ostriches as strong argument of possibility of such change,—as we see them in space, so might they in time" (1987a, pp. 70, 175). The question marks are gone, the syntax is regular though abbreviated—symptoms that private persuasion has been completed, that a belief has become a relatively stable component of the network of theoretical statements that constitutes Darwin's professional self.

There is a second theme of Darwin's speculation: the importance of "neutral ground" in the determination of good species. Around mid-May 1838, the ostriches become an example of a clearly formulated rule for species determination: "analogy from every country & class tells us that ¿O[petiorhynichus]. Modulator & O. Patagonicus. till neutral ground ascertained, call them varieties. but two ostriches good species because interlock [without interbreeding]."[18] An entry written a bit later suggests a necessary condition for such speciation: "ISOLATION of range . . . tends to alteration views.—ostriches d[itt]o—" (1987a, pp. 277, 304).

By mid-1838, then, only months before the Malthusian revelation, Darwin's speculations concerning the neutral ground shared by two ostriches have stabilized: variations produced by descent with modification and isolated by geographic barriers may form good species. These will not interbreed, even if, at some later time, their ranges become contiguous.

Only after a hiatus of twenty-two years, in the *Origin of Species*, are the ostriches presented again to the public. But in the *Origin* they are introduced not as the source of important theoretical conclusions but as illustrations of the geographic succession of closely allied species, a phenonomenon that can be best accounted for by descent with modification:

> The naturalist in travelling . . . from north to south never fails to be struck by the manner in which successive groups of beings, specifically distinct, yet clearly related, replace each other. He hears from closely allied, yet distinct kinds of birds, notes nearly similar, and sees their nests similarly constructed, but not quite alike, with eggs coloured in nearly the same manner. The plains near the Straits of Magellan are inhabited by one species of Rhea (American ostrich), and northward the plains of La Plata by another species of the same genus; and not by a true ostrich or emeu, like those found in Africa and Australia under the same latitude . . . We see in these facts some deep organic bond, prevailing throughout space and time, over the same areas of land and water, and independent of their physical conditions. The naturalist must feel little curiosity, who is not led to inquire what this bond is.
>
> This bond, on my theory, is simply inheritance . . . (1964, pp. 349–350)

The significant distinction in argument between the Notebooks and the *Origin* lies in the contrast between what convinced Darwin and what Darwin thought would persuade his first audience. This is the only explanation necessary for a range of differences in argument building, including the one just examined. This explanation applies also to the most striking difference in argumentative strategy between the Notebooks and the *Origin:* the fact that the analogy between artificial and natural selection, which looms so large in the latter, is of such trifling importance in the former.

Science as Problem Solving

In the Notebooks, as in the *Origin,* Darwin is concerned with the nature or *quale sit* of evolutionary change; in both he forms analogies and generalizations and supports these with facts or arguments. Between the Notebooks and the *Origin,* there is a difference in persuasive strategies, but there is no disparity in the nature of the reasoning Darwin employs. His thought resembles ordinary problem solving, not some special model of scientific thinking, such as the hypothetico-deductive method.[19] Those who claim otherwise of the Notebooks cite this passage: "The line of argument {often} pursued throughout my theory is to establish a point as a probability by induction & to apply it as hypothesis to other points. & see whether it will solve them." This certainly appears to be evidence that Darwin used the hypothetico-deductive method. But it is not; Darwin later testifies to a persistent difficulty in separating hypothesis formation from testing, a separation essential to the hypothetico-deductive method: "I from looking at all facts as inducing towards law of transmutation, cannot see the deductions which are possible" (1987a, pp. 370, 410).

Because the spirit of intellectual play governs all of the Notebook entries, it governs these on methodology: in the first passage, Darwin is attempting to fit his thinking into the pattern Whewell describes as uniquely scientific; in the second, he recognizes the serious limitations of his first formulation as a description of his thought. And so should we. The heart of the Notebooks is a "state of *indecisive flux.*"[20]

Darwin's thought also resembles ordinary problem solving in its conservatism. Summarizing the current consensus, Robert Richards speaks of Darwin's "persistence in retaining and modifying ideas rather than simply dropping and replacing them" (1987, p. 83). But neither Richards nor his predecessors point out that this conservatism is not a Darwinian peculiarity but a mental characteristic that must be general if we are to maintain a sense of self. Under any plausible construal, "there must be continuity between a person's beliefs one day and his beliefs the next."[21] Moreover, these beliefs cannot be altered at will. Although at first glance this claim seems counterintuitive, it simply articulates a logical truth. It is self-contradictory to say that I can change my beliefs about the world without reference to the world: "Only if I think of my beliefs as forced upon me by the facts will I believe them" (Swinburne 1984, p. 63). Indeed, this is the theme of the famous first paragraph of the *Origin:* "When on board H. M. S. 'Beagle,' as a naturalist, I was much struck with certain facts in the distribution of the inhabitants of South America, and in the geological relations of the present to the past inhabitants of that continent. These facts seemed to me to throw some light on the origin of species . . ."[22]

The general claim of conservativism apparently leaves unexplained the two cases in which Darwin testifies to sudden flashes of insight. Of natural selection he says: "it at once struck me"; of the link between diversity and niche capacity, he states: "I can remember the very spot in the road . . . when to my joy the solution occurred to me" (1958, pp. 120–121). But there is no conflict between the thesis of conservatism and the fact of insight. Insight is not a logical step that is larger than usual. It is not a logical step at all; it is a psychological state, an experience of sudden conceptual reconfiguration in the face of a pressing intellectual problem (see Köhler 1947, pp. 188–210). The suddenness with which we experience insight says nothing about the psychological and rational processes that underlie its possibility, processes that need not be sudden; *a fortiori*, the fact of insight cannot be evidence against the essential conservatism of personal systems of belief.

It may be objected that the account of Darwin's cognitive style leaves no room for his genius. But the isolated jottings of a genius need not demonstrate that genius; to assume so without evidence is to commit the fallacy of composition. It should not undermine our view of Darwin to find that the individual entries in his Notebooks, the individual moves from theory to theory and from fact to theory, show no evidence of the intellectual stature to which the Notebooks as a whole, and his life's work, bear unequivocal witness. The entries in the *Red Notebook* are "brilliant scraps" only in the retrospective light of Darwin's acknowledged genius.[23]

I claim in this chapter that Darwin's most creative phase is appropriately described as a rhetorical transaction within the self. The Notebooks enact a drama of self-persuasion: Darwin is driven forward by the rush of new concepts and facts, yet held in check by the need to maintain the self as a coherent network of beliefs. As the elements in Darwin's complete theory approach their final shape, his style alters from a means of expression apparently in close touch with primary mental processes to one that anticipates public forums.

Coincident with this stylistic process, the arrangement of the Notebooks alters in character: isolated facts and concepts coalesce, intertwine, and, finally, unite into more and more fundamental lawlike statements. But it is not in these statements that the unique character of the Notebooks lies; it is rather in the mental processes initiated in the most disjointed, least comprehensible entries. These most clearly display the special epistemic flavor of the Notebooks: a disciplined lack of commitment to the full truth of assertions, a deficiency that enables the evolutionary transformations to final theory.

PART III

Science and Society

CHAPTER 11

The Emergence of a Social Norm

In a series of masterful essays, Robert K. Merton develops the thesis that conflicts over priority in science are not, as had been assumed, the product of unfortunate personal aberrations; rather, they constitute a strategic research site for sociological analysis; indeed, they open a path to the paradox at the heart of the scientific enterprise, the paradox of communal competitiveness: although the general progress of scientific knowledge depends heavily on the relative subordination of individual efforts to communal goals, the career progress of scientists depends solely on the recognition of their individual efforts.

This paradox, and the priority conflicts it engenders, were not always a part of scientific activity; they are a historical phenomenon, one whose origin can be traced to England in the middle of the seventeenth century, a time when men deliberated about science and changed their minds concerning the social structure most appropriate to its advance. In the seventeenth century, one could say, the social norms of science changed.

Merton's work exemplifies the need for sociological analysis to identify the structural conditions that facilitate such changes. Nevertheless, norms themselves emerge and gain (or lose) support in contexts in which rhetorical analysis seems not only appropriate, but necessary. For Clifford Geertz, it is only through the interaction of social forces and rhetoric that "ideologies transform sentiment into significance and so make it socially available."[1] Working within rhetorical theory, Thomas Farrell describes this same process of transformation, one by which social consensus initially attributed by speakers to audiences produces social norms. To Farrell this emergence, this "ability of rhetorical transactions gradually to generate what they can initially only assume appears to possess a rather magical ambience" (1976, p. 11).

My purpose in this chapter is analyze this "ambience," to translate magic into rhetorical analysis. The case chosen is deliberately one already subjected to extensive sociological scrutiny. This reexamination, this parallel critical excursion, by making as clear as possible the contrast between rhetorical and sociological analyses, should highlight the value of the former in explaining social change.

The facts are not in dispute. At the beginning of the seventeenth century, scientists generally worked independently, disseminating the results of their efforts by means of correspondence and an occasional book. From time to time, a scientist might show concern for the order of discovery. But such concern was not an institutional norm. By the end of the century, the situation was radically different: concern for priority in discovery had become just such a norm, with prior journal publication its criterion.

Bacon's Cooperative Vision
of Science in *New Atlantis*

That this rhetorical story begins with a work of fiction necessitates some explanation of the scope of rhetoric: *New Atlantis* (1627) is a Utopian novel, not an argument. But for Aristotle rhetoric is not about argument only: it is about *all* the means of persuasion. *New Atlantis* is a novel with a difference: it makes the claim that a society is best when founded on the advance of science through cooperation; it persuades not through argument, but through its vivid presentation of the benefits of that society. Salomon's House is the intellectual center of the island kingdom of Bensalem. Its goals are lofty: "the knowledge of Causes and secret motions of things, and the enlarging of the bounds of Human Empire, to the effecting of all things possible" (Bacon 1937, p. 480). In pursuit of these goals, the members of Salomon's House perform many of the functions later entrusted to the Royal Society: the members hold meetings, communicate regularly with one another, and publish their results.

At Salomon's House science is entirely a cooperative activity, pursued according to a firm division of labor: experimenting and the making of inferences are, for example, entrusted to different groups of workers. The science of Salomon's House is, moreover, cooperative in a larger sense; it is an activity entirely free from any taint of nationalism or imperialism: "We maintain a trade . . . only for God's first creature, which was *Light*" (Bacon 1937, p. 469).[2] The reference to Deity is apposite: even science is a form of worship: "for the laws of nature are [God's] own laws" (p. 459). In Bensalem science is an activity totally in harmony with itself, with society at large, and with God's purposes, and it is in no way subordinate to the political purposes of the state (p. 489). Although the discovery of new knowledge is not a competitive activity—there is no competition anywhere in Bensalem—discoverers are singled out for honor and reward.

In *New Atlantis* Bacon uses metaphor to underline the high-mindedness of the Utopian vision that is his persuasive vehicle. He does this by varying the distance between tenor and vehicle in his metaphors, adjusting "the difficult and elusive relationship between A and B."[3] For example, throughout *New Atlantis,* Bacon uses light as a metaphor for intellectual and spiritual illumination. In this comparison there is no ideational distance between light and the creative act. It is as in Genesis: "God said, Let there be light, and there was light." But in this same work Bacon compares scientific communication (the tenor) to trade (the vehicle); in addition, he compares the increase in scientific knowledge (the tenor) to the advance of Empire (the vehicle). In the latter cases, however, he uses these comparisons only to increase the ideational distance between science and both trade and imperialism: the trade is in light, the Empire human (compare Jardine 1974, pp. 202–205).

Potential Conflict in Sprat's
History of the Royal Society

Before the founding of the Royal Society of London for the Improving of Natural Knowledge in 1662, Englishmen published numerous schemes for the advancement of science, schemes whose obvious parent was Salomon's House in *New Atlantis* (see Jones 1961, pp. 170–176 and 317). Sprat's *History of the Royal Society,* published just five years subsequent to the Society's founding, is in this tradition.

It is less the history than the manifesto of that organization; as such, it is a clear embodiment in argument of Baconian myth that sees national and transnational cooperation as the ideal truth-seeking procedure in the natural sciences (Sprat 1667, p. 99 ff.). Sprat notes "how much progress may be made by a form'd and Regular *Assembly*" that coordinates "the joynt force of many men" of all ranks, Englishmen who "*work* and *think* in company, and confer their help to each others *Inventions*" (pp. 28, 39, 427). These cooperative goals are not confined to England. Science prospers "when it becomes the care of united Nations" (p. 3). Cooperation so all-encompassing implies a free exchange of information across national borders: "the benefit of a universal *Correspondence*, and *Communication*" (p. 424).

But in the *History of the Royal Society* another voice is also heard, a voice at odds with the spirit of cooperation: "*Invention* is an *Heroic* thing, and plac'd above the reach of a low, and vulgar *Genius* . . . a thousand difficulties must be contemn'd, with which a mean heart would be broken . . . some irregularities, and excesses must be granted it, that would hardly be pardon'd by the severe *Rules of Prudence*" (p. 392). A view of invention so hyperbolic, a stance so obviously motivated by the "desire of glory" (p. 74), is incompatible with Utopian cooperation. Even more jarring to the spirit of cooperation is Sprat's transformation of discovery into a form of capital whose owner needs better protection from society: "while those that add some small matter to things begun, are usually inrich'd thereby; the *Discoverers* themselves have seldom found any other entertainment than contempt and impoverishment . . . The fruits of their *Studies* are frequently alienated from their Children" (p. 401). In Sprat's myth, in contrast to Bacon's, the heroic and the commercial are indissolubly linked: fellows will "reap the most *solid honor*"; but they "will also receive the strongest assurances, of still retaining the *greatest part of the profit*" (p. 75).

Sprat's other, competitive, voice also reaches beyond national borders: scientific advance is an integral component of England's imperialist destiny. The English character is best suited both to science and to command, indeed, to Empire (pp. 113–115, 420). England, therefore, "may justly lay claim, to be the Head of a *Philosophical league*, above all other Countries in *Europe*" (p. 113); London, moreover, is its ideal center: "the head of a *mighty Empire*, the greatest that ever commanded the *Ocean*" (p. 87). Imperialism and science work together: scientific advance can materially aid "the advancing of *Commerce,* as the best means . . . to enlarge their *Empire*" (p. 408): "if ever the *English* will attain to the *Mastery* of *Commerce,* not only in *discours,* but *reality:* they must begin it by their *labors,* as well as by their *swords*" (p. 423).

When Sprat is praising cooperation, he uses trade and imperialism in a Baconian fashion, as vehicles for metaphors that create ideational distance between these activities and natural science, an activity whose purpose is "*to increase the Powers of all Mankind, and to free them from the bondage of Errors . . . [a] greater Glory than to enlarge* Empire" (unnumbered Epistle Dedicatory):

> By their *naturalizing* Men of all Countries, [the Royal Society has] laid the beginnings of many great advantages for the future. For by this means, they will be able, to settle a *constant Intelligence,* throughout all civil Nations; and make the *Royal Society* the general *Banck,* and Free-port of the World: a policy, which whether it would hold good, in the *Trade* of *England,* I know not: but sure it will in the *Philosophy* [natural science]. (p. 64)

Indeed, in contrast with the imperialist wars of the times, the progress of science is depicted as a war of "all civil Nations" against "*Ignorance, and False Opinions*" (p. 57).[4] But when Sprat is waxing patriotic, matters are quite otherwise. England's command over the seas has as its purpose "to bring home matter for *new Sciences,* and to make the same proportion of Discoveries above others, in the *Intellectual* Globe, as they have done in the *Material*" (p. 86).

The aims mirrored in these passages are only potentially in conflict. There is no indication that Sprat—or anyone else at the time— noticed any incompatibility between the cooperative and transnational aims of the Society and the subordination of science to the national and imperial ambitions of England: no one, it seems, foresaw the oxymoronic potential of a *transnational* society founded under *royal* auspices. In all probability, for neither Sprat nor for the founders was this subordination anything more than another argument for the social support of science, one that was, at the same time, a laudable expression of an emerging patriotism. An author can refuse intellectual engagement, albeit at the price of some coherence;

but an actual social group simultaneously pursuing cooperation and competition, transnationalism and imperialism, is plainly on a collision course with itself.

The Newton-Leibniz Dispute

In the minds of its founders, and of its first Secretary, Henry Oldenburg, the Royal Society embodied the twin virtues of cooperation and transnationalism—the essence of Bacon's vision, the burden of Sprat's argument. In Oldenburg's words: "The object of science [is] of so vast an extent, that it demand[s] the united genius of more than one nation to exhaust the subject" (Birch 1968, I, 317). To Huygens of Holland, Oldenburg wrote that "there is no doubt at all that, if we press onwards at a steady pace, maintaining a frank and regular correspondence for our mutual benefit, we shall in time see considerable progress in every branch of science" (1968, V, p. 583). To René Sluse, the Flemish mathematician, he addressed the following hyberbolic peroration: "Throw your lot in with us, as the French and Italians have done. We earnestly pray that all your people, all Germany, and the whole of the Netherlands may associate their labors with ours" (III, 537–538). Throughout his voluminous correspondence, Oldenburg's indefatigable efforts in recruiting and encouraging all scientists, English and Continental, are everywhere apparent.[5] Indeed, so trusted on the Continent was the Society under the stewardship of its first Secretary—its last European Secretary— that Oldenburg was able to mediate, if not always to resolve, disputes, including disputes over priority, most of which were between Englishmen and their Continental counterparts.

Such disputes arose as a matter of course when the conflicting rhetorical aims in Sprat's *History* became actual social norms, real standards of behavior among the first Fellows and correspondents, men who simultaneously worked for and against each others' interests. Even during the first years of the Society, problems of scientific ownership appeared (1966, II, 291, 486). And, despite the ready availability of a register of discoveries, priority disputes broke out regularly: between Hooke and Newton; Hooke and Auzout; Huygens, and Wallis and Wren; Huygens and Wallis; van Heuraet and Neile; Huygens and Hooke; Wren and Auzout; Hooke and Mercator.[6]

Hooke and Newton, Wallis and Wren—these disputing pairs were

Englishmen. In the other disputes, however, the conflict was not only between scientists, but between England and the Continent. In a letter to Oldenburg, John Wallis, the English mathematician, disowned competitive motives on the part of his fellow countrymen: "Whatever may be thought of the French or the Dutch, certainly the English are not thus given to continual pursuit of fame" (Oldenburg, 1975, X, 4).[7] But his more usual views belied these Utopian sentiments: "Onely I could wish that those of our own Nation; were a little more forward than I find them generally to bee (especially the most considerable) in timely publishing their own Discoveries, & not let strangers [foreigners] reape ye glory of what those amongst ourselves are ye Authors."[8]

These international disputes reached their climax in the quarrel between Newton and Leibniz over priority in the discovery of the calculus. According to the social theory favored by Victor Turner, this quarrel is a social drama. Insofar as they participate in such a dispute, the actors in a social drama live not by ordinary but by dramatic time, and they exhibit not everyday but heightened responses (1982, pp. 9–10). Although social dramas have identifiable acts or stages, in a drama as complex and lengthy as the calculus dispute it would be futile to try to map these precisely onto the actual sequence of events (1978, p. 79). This lack of perfect correspondence, however, does not vitiate the theory as an explanation of the text with which we are now concerned, the "Account of the Book entitled *Commercium Epistolicum,*" an anonymous summary of the calculus dispute published in the *Transactions* of the Royal Society in 1715. The "Account" is clearly a form of redress, the stage at which the representatives of a society try to come to terms with the fundamental cleavages that have driven a particular social drama to crisis (1982, pp. 69–70, 78).

The calculus dispute came to a head in the second decade of the eighteenth century. In 1712, under the press of events, the President of the Royal Society convened an international committee to investigate the twenty-year controversy between Newton and Leibniz concerning priority in the invention of the calculus. Within fifty days they reported in Newton's favor, condemned Leibniz, and issued in evidence a collection of documents, the *Correspondence of John Collins and Others about the Development of Analysis,* usually known by its shortened Latin title, *Commercium Epistolicum.* In 1715 there appeared in the *Philosophical Transactions* of the Society a shortened version of the *Commercium,* the "Account."

Any attempt at redress may prove unsuccessful. So it was with the "Account," and with the inquiry that preceded it. The adjudicating body, and the documents issued in its wake, were so seriously compromised as to undermine credibility completely. The "Account" asserts that "no Man is a Witness in his own Cause. A Judge would be very unjust, and act contrary to the Laws of all Nations, who should admit any Man to be a Witness in his owne Cause" (Hall 1980, p. 284). But the sole author of the 'anonymous' "Account" was none other than Newton himself. Nor was this his worst. As President of the Royal Society, he had personally selected the committee of inquiry, carefully stage-managed its deliberations, and thoroughly supervised its report.

This brazen behavior had implications far wider than the calculus dispute. From the middle of the sixteenth century, which saw the collapse of the Church as the transnational State, there had been no legitimate governor of the conduct between nations. It is in the century of the Royal Society, the first full century to bear the brunt of this absence, that Grotius wrote his ground-breaking *De Jure Belli et Pacis* (1625) "to see whether there were not certain common duties generally felt as binding, if not always practiced, and to set forth an ideal" (Figgis 1960, p. 246). And it is in the context of this general need for international law that we should see Oldenburg's attempt to legitimate the Royal Society as a means of enforcing common duties in science—what we would call social norms. But Newton's outrageous conduct signaled the utter unraveling of Henry Oldenburg's efforts to transform the Royal Society into a transnational body capable of adjudicating scientific disputes.[9]

A failure at redressing the dispute between Leibniz and Newton, the "Account" was nevertheless successful at making vivid a vision of science diametrically opposed to that of Salomon's House, one which dramatizes the claim that science is wholly competitive and wholly national in character. The "Account" realizes this rhetorical vision by inventing two dramatis personae: Newton, who stands for honor, genius, England; Leibniz, who represents immorality, intellectual theft, Continental intrigue. Although the "Account" persuades us neither of Leibniz's guilt nor Newton's innocence, by its twin characterizations it convinces us that the differences between these men are indeed irreconcilable.

By casting the "Account" in the form of a prosecutor's summation before a jury, Newton makes this irreconcilability a given of his case: unrelenting verbal attack becomes not a pleasure to indulge but a duty to discharge. In the following lengthy passage—only length can adequately display his forensic skills—Newton combines historical evidence with sarcasm in a devastating assault on Leibniz's credibility:

> When Mr. *Newton* had received this Letter, he wrote back that all the said four [mathematical] Series had been communicated by him to Mr. *Leibnitz* . . . Whereupon Mr. *Leibnitz* desisted from his Claim. Mr. *Newton* also in the same Letter dated *Octob*. 24. 1676. further explained his Methods of Regression, as Mr. *Leibnitz* had desired. And Mr. *Leibnitz* in his Letter of *June* 21. 1677. desired a further Explication: but soon after, upon reading Mr. *Newton's* Letter a second time, he wrote back *July* 12. 1677. that he now understood what he wanted; and found by his old Papers that he had formerly used one of Mr. *Newton's* Methods of Regression, but in the Example which he had then by chance made use of, there being produced nothing elegant, he had, out of his usual Impatience, neglected to use it any further. He had therefore several direct Series, and by consequence a Method of finding them, before he invented and forgot the inverse Method. And if he had searched his old Papers diligently, he might have found this Method also there; but having forgot his own Methods he wrote for Mr. *Newton's*. (Hall 1980, p. 279)

Working among the Ndembu, Victor Turner discovered that their social dramas tended to widen until they coincided with their society's deepest conflicts; that crisis tended to display, and redress inscribe, those conflicts (1982, p. 70). It is this tendency that accounts for Newton's view of the priority dispute as an element in the struggle between nations. In English, in an English journal, Newton plainly addresses an audience solely of Englishmen, an audience viewed as both judge and jury (Hall 1980, pp. 309, 310, 313).[10]

This tendency for social dramas to widen and deepen accounts also for Newton's conviction that the calculus dispute represents a difference over ultimate issues. Leibniz is wrong, not only as a man and a mathematician, but in every possible way and at every possible level: "It must be allowed that these two Gentlemen differ very much in Philosophy [methods of doing science]. The one proceeds upon the Evidence arising from Experiments and Phaenomena, and stops where such Evidence is wanting; the other is taken up with Hypotheses, and propounds them, not by Experiments, but to be believed without Examination." Indeed, Leibniz is mistaken even about the role of the Almighty in the conduct of the physical world (Hall 1980, p. 314)!

It is another characteristic of social drama that it eventually asserts its independence from its main disputants. In the case of the calculus dispute, the flames of controversy were just as likely to be fanned by others as by Leibniz or Newton; and the quarrel survived Leibniz's death. As it advanced, it deepened; as it deepened, it darkened: the motives of the opposition became automatically suspect, and all attempts at mediation failed. By the time of Newton's "Account," the dispute had been transformed into an irresolvable debate over irreconcilable ideological views, a debate that promised to renew itself in virulence any time two scientists disagreed about priority.

This tendency of the contestants in priority disputes to lose control made imperative an administratively simple solution that would reduce the frequency of such disputes and contain within the bounds of normal civility those that actually did break out. But before we carry the story of priority to this next stage, the stage at which the institutional response approaches adequacy, we must pause to scrutinize more carefully two assumptions that Newton was careful not to scrutinize at all—that disputes over priority are disputes over intellectual property, and that they are national in character.[11]

Priority as Property

What was at stake in the calculus dispute was the ownership in perpetuity of this branch of mathematics: exclusively the invention of one Englishman, exclusively the product of the English genius for science. This concern for the scientific ownership that priority affirms is wholly a historical phenomenon. Science is not necessarily conscious of priority; the norms of modern science are not the norms of the ancient Greek or the medieval enterprises of that name. Not until the middle of the seventeenth century did a concern for scientific originality take roughly its present form.[12]

In fact, so persuasive is this form that it seems natural today to speak of the *product* of scientific activity as, in some sense, the *exclusive property* of a *person:* why? In the seventeenth century, at least two notions of scientific property were broached. In Huygens's opinion, credit for a scientific discovery "ought to be equally assigned to all those who find out a thing without regard to the time [when it was done], provided that they can state firmly that they have made the discovery without any assistance" (Oldenburg 1968, V, 362). In his controversy with Leibniz over the calculus, Newton asserts his very different point of view: "second Inventors have no Right. The sole Right is in the first Inventor until another finds out the same thing apart. In which case to take away the Right of the first Inventor, and divide it between him and that other, would be an Act of Injustice" (Hall 1980, p. 305; see also p. 308).

Neither view of scientific property is explicable without recourse to the revolution in the law of property that accompanied the rise of capitalism. New forms of enterprise, such as joint stock companies, required new law; at the same time, the commercialization of society altered the legal status of existing endeavors. From time immemorial, a plow or a book were forms of property, chattels; now, for the first time, the *idea* for a plow, or for a book, were also forms of property—very different forms, new to the law. Patents and copyrights gave inventors and authors a kind of monopoly: the rights to manufacture and to print could be exclusively owned; as property, they could be bought and sold.[13]

Both Newton and Huygens claim that priority establishes a right to intangible property of this new sort. It is, of course, not precisely a legal right. The right to priority in scientific discovery cannot be licensed, reassigned, or sold. But priority is precisely analogous to a property right: its proper assertion creates in every scientist the duty to acknowledge another's ownership (Hart 1980, pp. 257–258).

Huygens's view of a common right did not prevail. Its source was in older ideas of property, ideas in the process of being superseded; for example, the feudal notion of the shared rights of tenants to a commons was already endangered and would not survive the eighteenth century. Newton's view, on the other hand, did survive. It clearly rode the crest of the new law of the Stuarts: the exclusive ownership of inventions and literary works was guaranteed, in the first case, by the exempting clause of the Statute of Monopolies, and, in the second, by the Copyright Act (Jenks 1949, pp. 284–285, 289). Under the sway of such exclusivity, only collaborators, and those making a precisely simultaneous independent discovery, could share a priority right.[14]

But why did priority become a national as well as an individual property right? On reflection, we need not be surprised by the close connection between capitalism, nationalism, and its twin, imperialism. Within England, the mid-century saw both agriculture and manufacturing reorganized on competitive lines in accordance with the nascent capitalism on which the industrial revolution would eventually be built. Internationally, in its search for raw materials and markets, England was involved in a bewildering succession of wars and alliances whose only common denominator was international competition, the imperial aim shared by England and the Royal Society. At the end of this period, signaled by the Treaty of Utrecht in 1713, England's colonial rivals, France and the Netherlands, were exhausted by a half-century of warfare. England, on the other hand, was invigorated by fifty years of national and commercial triumph, a vigor manifested very evidently in the rise of English science under the aegis of the Royal Society. It is no accident that in the "Account" Newton's forensic skills are animated by a set of values that were also the basis of the British Empire.

Dealing with Priority Disputes

If the possibility of strife over priority were not to be a continuous threat, the establishment of an administratively simple means for determining priority was imperative. This means had to be persuasive; that is, it had to be acceptable to the community of scientists in the absence of a transnational adjudicating body, an International Court of Scientific Disputes. The means the Royal Society eventually adopted is still in place. Scientific discoveries became, and have remained, exclusive properties, held by a right analogous to copyright, a right at the same time personal and national; their priority was, and is, routinely guaranteed by dated journal publication, a procedure that simultaneously fixes the ownership necessary for competitive advantage, and guarantees the broadcast of ideas demanded by the imperative to share. Since it simultaneously satisfies two potentially conflicting social norms, dated journal publication seems a natural and elegant resolution of the paradox of communal competitiveness (compare Zuckerman and Merton 1973, p. 465).

But matters were far less clear for scientists in the seventeenth century. In the first place, the distinction between printing and publishing was very real at the time. Publication was making public; it was the sharing of ideas. Printing was only one means to this end. Writing to Oldenburg in 1665, Adrien Auzout stated the general view: "I see little difference between printing scientific matters contained in letters and showing these same letters to those learned in these matters who can copy them out when they have them on loan" (Oldenburg 1966, II, 518). Speaking of a scientific discovery, Viscount Brounckner said that its author "did then communicate & publish ye same (though not in print) to my self & others" (Oldenburg 1975, X, 291). Giving a lecture was also a means of publication: Wren regarded a lecture as "publication enough" (Birch 1968, I, 48). From this point of view, lectures and letters were the equivalent of dated journal publication. But by the second half of the century, such notions were becoming obsolete; indeed, these last two clarifications are embedded in disputes over priority.

Seventeenth-century scientists were undecided not only on what publication meant, but on whether the broadcast of completed work was the best means of securing priority. Huygens suggested an alternative to full publication: the announcement of work in progess in anagrammatic form (Oldenburg 1968, V, 556). Oldenburg suggested an alternative to anagrams; he was confident that the temporal priority of scientific discoveries could be established by means of the registration stipulated in the founding documents of the Society (Weld 1858, p. 527). He constantly advertised this option and himself used the register to resolve a dispute.[15] In 1665 he proposed to regularize the process: "Mr. OLDENBURG made a motion in the name of some member of the society, that when any fellow should have a philosophical notion or invention, not yet made out, and desire, that the same sealed up in a box might be deposited with one of the secretaries, till it could be perfected, and so brought to light, this might be allowed for the better securing inventions to their authors" (Birch 1968, II, 24; see also p. 212).

But registration was not without problems. The register, of which Oldenburg was so proud, was manifestly incomplete.[16] And even though the Society accepted Oldenburg's proposal, they worried that their fellows might be tempted to register projects that were far from completion. In an amendment, they insisted that any invention so registered be perfected "after about a year's interval" (Birch 1968, II, 25).

By the seventies, the tide had turned permanently in favor of

securing priority by the simple expedient of publication in the newly
founded *Philosophical Transactions* (Oldenburg 1975, X, 67). In the
last twenty years of the century, the trend toward regular publication
became marked. In 1683 the Society passed a resolution to emphasize
the importance of record keeping and putting things in writing (Birch
1968, IV, 251). By 1686 there were clear signs of a regular procedure:
Papin's paper was read, registered, and published, as were papers by
Halley, Hooke, Vossius, and DeVaux.[17] The trend toward dated
journal publication as a guarantee of temporal priority soon took full
hold: by publishing in print accounts of their completed discoveries
or inventions, scientists simultaneously shared their results and established their claim to these results as theirs alone. By creating the
means by which the ownership rights to its knowledge could be
established, the scientific community transformed rhetoric into social
reality.

The Implications of a Social Invention

If the story of priority ended here, it might mislead; it might imply
that the consequences of basing the reward system of science on the
respective dates of receipt of rival journal articles were uniformly
positive. But a focus on priority distorts: it misrepresents the history
of science and exaggerates the importance of originality in scientific
advance.

   In any important case, the real question of priority is likely to be
beyond settlement. George Stigler, a Nobel laureate in economics,
asserts that "when an idea is more than a technical definition or a
highly specific analysis, the temporal priority is hopelessly obscure.
Who first discovered [such major ideas in economics as] diminishing
marginal utility, or diminishing returns, or the quantity theory, or
the theory of underemployment equilibrium? I do not know, but it
is common knowledge that all such ideas have long histories before
they are stated by the men who made them important" (1965, p. 3).
Thomas Kuhn's classical analysis, "Energy Conservation as an
Example of Simultaneous Discovery," gives us a detailed example of
this problem in physics. Kuhn concludes: "What we see in [the works
of these scientists] is not really the simultaneous discovery of energy
conservation. Rather it is the rapid and disorderly emergence of the

experimental and conceptual elements from which that theory was shortly to be compounded" (1977, p. 72)

A focus on priority also emphasizes originality to the exclusion of other factors vital to science. The testing of hypotheses, the accumulation of knowledge in a particular field, the refinement and elaboration of a theory—these are three important components of scientific advance in which originality is, at best, a minor factor (Stigler 1965). Moreover, discoveries as great as the calculus and natural selection were multiple and nearly simultaneous. Had Newton and Darwin never lived, the calculus would have been discovered by the end of the seventeenth century, natural selection by the middle of the nineteenth. How important is the originality that priority certifies when such discoveries are virtually inevitable?[18] Furthermore, even scientists to whom considerable originality is usually ascribed function for the most part in a more ordinary capacity: even Newton and Darwin spent most of their scientific lives working out the implications of their great insights.

Indeed, a concentration on priority entails an emphasis on originality that may actually impede scientific advance: "Quite aside from the fact that much original work is mistaken, an excessive rate of production of original work may retard scientific progress . . . When the rate of output of original work gets too large, theories are not properly aged. They are rejected without extracting their residue of truth, or they are accepted before their content is tidied up and their range of applicability ascertained with tolerable correctness. A cumulative slovenliness results" (Stigler 1965, p. 14).

Finally, too great a concern with priority distorts the efforts of scientists; it encourages them to show a "concern with recognition" at the expense of a "concern with advancing knowledge" (Merton 1973, p. 338): in other words, the actual effect of an emphasis on priority may undermine its intended effect, the encouragement of scientific advance. Every scientific paper instantiates the tension created by these opposing tendencies.

If my analysis is correct, rhetoric is an essential component in social change, and rhetorical analysis is an essential ingredient in sociological analysis. But the complementary nature of rhetorical and sociological analyses may indicate a deeper kinship. Writing this chapter, I was struck by the ease with which I selected priority disputes as a strategic research site for rhetorical analysis. No rhetorician had gone over this ground before; but surely no one studying the rhetoric of early modern science could ignore this forensic Everest. By way of contrast, one may note the initial resistance among responsible scholars to Robert Merton's recognition of these same priority disputes as a strategic site for sociological research, a resistance Merton points to with the legitimate pride of a pioneer. He quotes his fellow scholar, Charles Gillispie, writing in 1958: "In a note . . . to Merton, I wrote that, although it seemed surprising that the phenomenon was so nearly universal an accompaniment to scientific discovery, I did wonder whether the matter wasn't a bit trivial; I don't believe I also said 'unworthy' but recollect that such a dark thought was in my mind" (quoted in Merton 1987, p. 22).

In addition, I was surprised by the ease of fit between Merton's central insight and my rhetorical analysis. In part, the source of this ease is Merton's own awareness of rhetoric as a category of sociological evidence. In his paper on priority, for example, he claims that "property rights in science [consist in] the recognition by others of the scientist's distinctive part in having brought the result into being." As support for this claim, he cites texts employing metaphors in which priority is the tenor, property the vehicle: for example, "Ramsay . . . asks Rayleigh's '*permission* to look into atmospheric nitrogen'" (1973, pp. 294–295; my emphasis). That this quotation occurs not in Merton's text but in a footnote is clear rhetorical evidence that, for him, sociology was the master, rhetoric the servant. Still, in my analysis of the rhetoric of Bacon, Sprat, Oldenburg, and Newton, I found so clear and complete, so serendipitous, a confirmation of the essential correctness of Merton's central thesis that more than mere confirmation seemed involved. Indeed, rhetorical analysis seemed unerring in its ability to penetrate the collective psyche of these pioneers of the social institution of modern science: patterns of figurative language in Sprat, for example, displayed Merton's conflicting imperatives at their birth, a manifestation of which, apparently, neither Sprat nor three centuries of readers had been aware.

Since Merton's conclusions were available to me, my rhetorical analysis cannot have the epistemological status of his groundbreaking work. My central insight concerning priority, an insight that organizes this chapter, is borrowed from sociology, not earned by rhetoric. But so serendipitous a relationship between sociological and rhetorical analysis may be an instance of a more general kinship between two allied disciplinary matrices. Perhaps neither has genuine epistemological priority; perhaps social forces, like capitalism and imperialism, forces that seem so impersonal, so like nature itself, are at bottom no less rhetorically constituted than is a concern for priority in scientific discovery. If that were so, the division of labor between rhetoric and sociology would still hold—sociology would still deal with the structural determinants of social conditions, rhetoric with their symbolic content and style—but for methodological, not epistemological, reasons.

CHAPTER 12

The Social Drama of Recombinant DNA

The routines of any society—building a bridge, processing an insurance claim—are designed to avoid rather than to encourage confrontation. But confrontation cannot be avoided entirely. On occasion, deep underlying conflicts, ordinarily "overlaid by the customs and habits of daily intercourse," manifest themselves "in public episodes of tensional irruption" (V. Turner 1974, pp. 33, 35). Victor Turner calls these episodes social dramas, periods in which society domesticates "the raw energies of conflict . . . into the service of social order" (1967, p. 39).

Social dramas move from threat to resolution; whatever the outcome of a particular conflict, cohesion is normally maintained: whoever wins, society is not the loser. By means of its social dramas, then, society attempts to turn public controversy into a reaffirmation of existing values. But a failed social drama is also possible. In its enactment, conflicting ideologies compete for hegemony at the expense of social cohesion.[1] The recombinant DNA controversy is such a drama.

Social dramas have four phases. The first, the breach, is an act deliberately defiant of social routines; it brings an underlying social conflict "into frightening prominence." If not sealed off quickly, the breach will widen into the second phase of the social drama—crisis, a state "coextensive with some dominant cleavage in the widest set of revelant social relations to which the conflicting or antagonistic parties belong" (V. Turner 1974, p. 35). Crisis is followed in turn by redressive action: society adjudicates rival claims by means ranging from informal mediation to formal justice before a supreme judicial body. Although redressive action is designed to resolve crises, in some instances it may have only a superficial result, one that does not lead

to "peace among contending groups." Consequently, "endemic, pervasive, smoldering factionalism" may remain to undermine the fourth
and last phase of social drama: reintegration. During this stage the
warring parties become part of a new status quo; failing that, their
antagonism is socially recognized and legitimated. Each phase of
social drama is distinct not only in function, but in other ways: "Each
phase has its own speech form and styles, its own rhetoric, its own
kinds of nonverbal languages and symbolisms" (V. Turner 1974, pp.
41, 43).

Turner's point is not that all societies are fundamentally alike;
rather, he holds that issues mobilizing fundamental conflicts in any
society inevitably do so in the form of a social drama. Each society
differs in its underlying conflicts. What becomes a social drama in
American society—an enactment of one of its deep underlying
conflicts—may be shunted aside as peripheral in British society. Or
it may become, in Soviet society, a social drama enacting a fundamental conflict more germane to Marxism.

From its beginning, the fiery American public debate over recombinant DNA was a parochial affair. Global interest remained generally
mild. Even in England, the only country where guidelines hardened
into law, "the debate [was] much more low key—so low key as to be
almost inaudible . . . Some unions [took up] the challenge, mainly on
the question of laboratory safety for individuals. But there were few
public meetings" (Morgan and Whelan 1979, p. 275). Soviet concern
about recombinant DNA, equally low key, had as its motive the desire
of biologists to avoid another Lysenko affair, another distortion of
science by politics. The American concern was deliberately misrep-
resented in the Soviet Union: according to Soviet scientists, the
research itself was not risky, though knowledge of recombination
could be abused. Soviet recombinant research continued alongside
an academic debate, a debate limited to the impact of recombination
on Marxism (Graham 1980).

Conflicting Ideologies

In the recombinant DNA controversy, two rival ideologies emerged:
technical and social. The technical ideology, eventually espoused by
the proponents of recombinant research, has its roots in the seventeenth-century defense of experimental science: "to plead for a

freedom for such as labour in the secrets of Nature, and that no discouragement be given such laborious and ingenious Inquirers." Such freedom is necessary because experimental science tarries for no one: "There is an infinite desire of knowledge broken forth in the world, and men may as well hope to stop the tide."[2] This union of science and nature is inevitably beneficial: "The Beautiful Bosom of *Nature* will be Expos'd to our view: we shall enter into its *Garden,* and tast of its *Fruits,* and satisfy our selves with its *plenty*" (Sprat; cited in Jones 1961, pp. 206–207, 227).

By the end of the nineteenth century this early enthusiasm concerning experimental science had in many cases turned to scientism, the view that natural science is the sole source of knowledge, including all ethical knowledge (see Habermas 1971, pp. 4, 67, 71; Frank Miller Turner 1974). Because "the social effects of science *must* be beneficial in the long run," the scientist could construe "events considered highly unlikely as actually impossible."[3] He need not consider himself "as a participant in a wider context of motives"; he need only "apply himself to his task as effectively as possible" (Burke 1969, pp. 30–31; emphasis deleted). What ethical issues arise may be treated as exercises in risk assessment, moves in a campaign to continue research, a campaign that systematically slights its possible social costs.

According to the social ideology espoused by the opponents of recombination, scientific advance is far from unproblematic. The purposes of science are not necessarily beneficial, and ordinary people, not just scientists, need to participate in scientific decisions that have a presumptive social impact. Like the technical ideology, the social ideology draws its strength from early roots. If, in the technical ideology, science is a symbolic promise of Eden, in the social ideology it is a symbolic threat of the Fall. Scientists must take their places in a line that stretches from Adam to Lex Luthor. Like Dr. Faustus, they "wonder at unlawful things." Nonetheless, with the unconcern of Dr. Frankenstein, they "[penetrate] into [their] causes," not realizing "how dangerous is the acquirement of knowledge." As a result of their efforts, like Dr. Rappaccini, they create new species by "adultery," "the production . . . no longer of God's making, but the monstrous offspring of man's depraved fancy, glowing with only an evil mockery of beauty."[4] A sense of moral superiority pervades the social ideology, a superiority no more substantial than the unquestioned optimism of the technical ideology. A vague populism, frequently combined with an unjustified suspicion of science and scientists, fuels a campaign designed to curb research, a program that systematically slights its possible social benefits.

Phases of the Recombinant DNA Drama

The Breach

The social drama of recombinant DNA was initiated in America not by the enemies of science but by scientists themselves. The breach was created by a public letter—the Berg letter of 1974—a document characteristic of the rhetoric appropriate to social drama's first phase. It was informed by "altruistic" intent and the belief of its ten signatories that they were acting "on behalf of other parties" (V. Turner 1974, p. 38). At the time, and subsequently, the ten signatories were much praised for the social responsibility reflected in the letter. Deeply stirred by their convictions, they felt "a sense of urgency about rapid publication" (Morgan and Whelan 1979, p. 283).

Their letter was quickly printed in *Science* and *Nature,* two of the most prestigious and widely circulated journals in science. It began on a scientific note but ended by recommending an ethical and political program: "Recent advances in techniques for the isolation and rejoining of segments of DNA now permit construction of biologically active DNA molecules in vitro [in an artificial environment]." But the process of recombination leads to "the creation of novel types of infectious DNA elements whose properties cannot be completely predicted in advance." Because these recombinations "could prove biologically hazardous" and have "unpredictable effects," the letter urged "all scientists working in this area to join" in a voluntary "deferring" of certain types of recombinant DNA experiments—a moratorium. It also recommended a conference, which later took place at Asilomar, where these matters could be discussed in detail (Watson and Tooze 1981, p. 11; see also pp. 5–6).

The Berg letter deals primarily with social knowledge, the "affiliative relationship of experiential beings," and only secondarily with technical knowledge, "the law-governed relationships among depersonalized phenomena" (Farrell 1978, p. 334). Nonetheless a social decision was made as a technical decision: by scientific criteria and by scientists alone. They were not merely to assess the risks of recombination; they were to ascertain whether those risks were to be borne by a public that was a party to neither determination. Asilomar was envisioned as a meeting of scientists, self-authorized as moral agents, acting for the public at large. The public was only a backdrop, a passive background; the scientists felt they had no need to check back with those they "represented."

Scientists at the Asilomar meeting confirmed the moral position of the Berg letter. The principal result was a fine-tuning of the moratorium recommendations: there were now four levels of physical containment for various recombinant experiments and four levels of experiment ranked in order of increasing risk (Watson and Tooze 1981, pp. 44–47). Nevertheless, some conferees seriously questioned the wisdom of the Berg letter; its stance on social responsibility was attacked as an essential abdication of trained intelligence. One of the letter's signers, James Watson, became openly and abusively critical:

> "Why . . . is *Xenopus* [frog] DNA safer to work with than, say, cow DNA?" . . .
> Paul Berg [the letter's originator] stood to get the session back on course: "We have to make a decision," he said. "Can we measure the risks numerically?"
> Watson, *sotto voce*, exploded, "We can't even *measure* the fucking risks!" (Rogers 1977, p. 75)

Watson's outburst stemmed not from the concern that led to Asilomar but from a purely technical ideology, which defines moral issues in terms of measurement, risk assessment, and scientific evidence. Those who espoused this ideology emphasized that the gradations of experimental danger had no scientific basis whatever. No one had ever been harmed by experiments with recombinant DNA, a record that would have surprised nobody, least of all the molecular biologists, had they thought seriously about risk assessment and considered that "if 10 scientists in each of 100 laboratories carried out 100 experiments per year, the least serious [most likely] accident . . . would occur on an average once in a million years" (Watson and Tooze 1981, p. 217). Such risks were evidently far lower than many of those of everyday life and especially of everyday industrial life, where "when you do engineering if you spend a million pounds you are going to kill somebody" (Morgan and Whelan 1979, p. 298).

The Crisis

Although Asilomar was at least partly the result of a moral impulse, the conference was carefully managed by its organizers: reporters were admitted only reluctantly, activist groups were severely excluded, and no record of proceedings was published. If the organizers believed that such behavior would protect the compromise Asilomar represented, a compromise that allowed recombinant research to continue under fairly stringent but self-imposed rules, they were mistaken.

Soon after the conference a period of crisis occurred. Its initiation was signaled by an open letter sent to all Asilomar participants from the Genetic Engineering Group of Science for the People. Signed by scientists from both Harvard and the Massachusetts Institute of Technology, the letter called recombination a "crossroad of biological research," citing the potential of recombination for "linking together of DNA molecules across natural species barriers." This linkage might have serious negative consequences, the letter said: perhaps "tragic results already caused by, for example, radium, asbestos, thalidomide, vinyl chloride and dieldrin"; perhaps "human genetic manipulation." The letter charged that such research was the result of a conspiracy of selfish interests that diverted attention from "the massive health needs of the population as a whole." The only way to reveal these motivations and to avoid dire consequences was to involve "the general public directly . . . in the decision making process" (Watson and Tooze 1981, p. 49). The letter's purpose was obviously political; it was an attempt by a group excluded from Asilomar to widen the scope of the conflict in order "to make certain that the power ratio among private interests most immediately involved shall not prevail" (Schattschneider 1975, p. 37; emphasis deleted).

All scientists opposed to the continuance of recombinant research shared a symbolic language. To Dr. Robert Sinsheimer of the California Institute of Technology, the molecular geneticist was an unwelcome interloper in the natural scene: "In a larger sense we have need to protect not only ourselves but the entire biosphere on which we depend and which is, in a sense, increasingly in our trust. We inherited and evolved in a marvelously balanced, self-sustaining world of life. Can we in truth predict what disruptions we may introduce into that world with our extraordinary inventions, our biological innovations, not derived from the historic evolutionary processes?" (Watson and Tooze 1981, p. 219). To Columbia University biochemist Erwin Chargaff, molecular biologists were guilty of megalomania: "Have we the right to counteract, irreversibly, the evolutionary wisdom of millions of years, in order to satisfy the ambition and the curiosity of a few scientists?" (1976, p. 940). To George Wald, Harvard Nobel laureate, such scientists were perpetrating a rape of nature: "I myself will do everything I can as long as I can to press for a deep principle in law of the inviolability of the human germ plasm" (National Academy of Sciences 1977, p. 218).

Nonscientists opposed to recombinant research shared this same ideological-rhetorical world. Francine Simring of the Friends of the Earth compared the recombinant DNA dispute to the "nuclear energy controvers[y]." Concerned that as a result of "human fallibility and technical failures" an accident would "inevitably occur," she wanted to "take into account the biohazards of accidental release of uncontrollable new organisms, the implication of interference with evolution" (Watson and Tooze 1981, p. 95). The mayor of Cambridge, Massachusetts, feared that molecular biologists "may come up with a disease that can't be cured—even a monster! Is this the answer to Dr. Frankenstein's dream?" (Rogers 1977, pp. 109–110).

In response to this moral assault, the majority of scientists, originally split on the issue of social responsibility, united behind the technical ideology, a set of beliefs that emphasized freedom of scientific inquiry without qualification, the practical benefits of research, and, by inference, the irrationality of their opponents.[5] What had begun at Asilomar as an idiosyncratic dissent embodied in one man, James Watson, an original signatory of the Berg letter, became the consensus ideology of the proponents of recombinant research. Watson's position, originally designed to undermine the compromise that led to federal regulation, became, with polarization, the ideological backing of any compromise that would allow recombinant research to continue.

At the National Academy Forum in March 1977, Stanley Cohen of Stanford University responded to a comment by Erwin Chargaff: "This statement and others suggest that you have for some years been deploring the search by man for scientific knowledge to dispel the mysteries of life." In response to a statement by Robert Sinsheimer, Paul Berg, also of Stanford, replied: "Dr. Sinsheimer has said, 'What if,' and you can go on with 'what ifs' eternally, and there is no way to answer all possible 'what ifs.'" The creed that energizes these responses was neatly expressed by Tracy M. Sonneborn of Indiana University: "It is no good, and unworthy of scientists, to impugn motives or run down opponents personally. The purely scientific issues can be scientifically settled only by facts; or, when facts are lacking, by getting them if possible; and if not possible, by rule of reason" (National Academy of Sciences 1977, pp. 56, 84, 287).

As a result of Asilomar, federal guidelines were formulated for the conduct of recombinant research. During the crisis stage of the controversy, the proponents of the research generally supported these guidelines as a shield for ongoing activities, while their opponents fought for legislation and court rulings to restrict further or to eliminate recombinant research. This confrontation of opposing ideologies was played out in numerous arenas: in books, in various periodicals, in hearings, in public forums (see Watson and Tooze 1981; compare Grobstein 1979 and Lear 1978). For the opponents, each arena represented a new audience, a new opportunity to widen the scope of the conflict, "to make certain that the power ratio among private interests most immediately involved shall not prevail." For these opponents, each new occasion represented a legitimation of the reality of the risk of research; if there were no risk, why were these meetings called, these hearings convened? As time went on, the proponents felt more and more on the defensive.

In late 1977 the issue came to a head: Congress was seriously considering legislation, laboratories on both coasts were alleged to have violated existing federal guidelines, and a court injunction had been sought against a proposed recombinant experiment (Watson and Tooze 1981, pp. 251–297). In retrospect, the proponents of recombination referred to the National Academy of Sciences Forum of that year as "probably our nadir point. Structurally it was set up to make it appear that the scientists were split 50–50" (Watson and Tooze 1981, p. 258)—a reluctant acknowledgment of the interim success of the opponents of research. After the forum the rhetoric of the proponents was less restrained, presumably an expression of prolonged frustration. James Watson, more outspoken than most, divided his opponents into three distinct categories: "kooks," "incompetents," and "shits" ("Recombinant DNA Research" 1977, p. 26). Polarization could hardly be carried further. The stage was set for the third phase of the social drama.

Redressive Action

In the crisis phase, the opponents of recombinant research wrested the issue from the control of the scientific community and successfully brought their case into the relatively uncontrolled arena of public debate. It was not in this arena, however, that the issue would be decided; it was in the courts, including the Supreme Court, in local legislatures, and in Congress itself—arenas expressly dedicated to social closure. In point of fact, neither the courts nor the legislatures fulfilled the hopes of the opponents of research; but neither were the fears of its proponents realized. The courts found for the proponents; nevertheless, they did not wholeheartedly espouse the technical ideology. When the District Court rendered its decision, it spoke in the accents of that ideology: "The experiment poses no substantial risk to human health or to the environment because (1) there is little likelihood the materials will escape from the maximum containment of the facility; (2) if such an escape did occur, the recombinant DNA molecules would not survive but would self destruct outside the laboratory environment; and (3) the particular virus being used has never been implicated in human disease" (Watson and Tooze 1981, p. 297). According to the Supreme Court, experiment itself was constitutionally protected by implication: "Legislative or judicial fiat . . . will not deter the scientific mind from probing into the unknown any more than Canute could command the tides."[6] But this was the view of a bare majority. Four justices supported another view: "It is the role of Congress, not this Court, to broaden or narrow the reach of the patent laws [to include or exclude newly created microorganisms]. This is especially true where, as here, the composition sought to be patented uniquely implicates matters of public concern" (Watson and Tooze 1981, p. 510).

It was in the legislatures rather than the courts that the opponents of recombinant research might have expected their success. Unlike the courts, legislative bodies are designed to respond to public pressure of a kind the opponents had successfully generated. But the balanced structure of legislative hearings led to a stalemate: lay persons were exposed at length to conflicting expert testimony by rival scientists and administrators, testimony whose conclusions they were in no position to evaluate fully. Moreover, as time passed, the case-clinching biological disaster failed to materialize; this circumstance, coupled with the protracted nature of the debate, gradually lowered the energy level within the legislative process until the priority for recombinant legislation was reduced below the threshold of possible enactment.

Finally, the proponents of recombinant research, while maintaining the force and tone of their public rhetoric, at last imitated their opponents by using the political process in their own interest: they opened a back channel to influence policymakers. Writing to a sympathetic colleague, Norton Zindler of the Rockefeller University illustrates the dialectic behind discourse whose purpose is influence rather than rational agreement: "As I told Paul [Berg] over the phone, I've been busy so long, calculating the results of moves—did I push too soon? too late? were the right people contacted? will he be angry at the truth? how far can I stretch the 'truth' without lying?" (Watson and Tooze 1981, p. 259).

Taken together, these events forestalled legislation. Now it was the opponents' turn to erupt in rhetorical fury and frustration at "'hard ball' politics": Pamela Lippe of the Friends of the Earth sneered at "the new data miraculously discovered and/or developed in only one year's time." These data offered only limited assurance of safety, an assurance especially unconvincing at a time when new and perhaps more dangerous experiments were being considered for approval under existing federal guidelines. Moreover, the situation promised to get further out of hand: "What little we know now will become even less as we continue to open broad n[e]w biological and ecological vistas" (Watson and Tooze 1981, p. 378).

Reintegration

Victor Turner offers alternate versions of reintegration, the last phase of social drama. The healthier involves the incorporation of warring groups and their conflicting ideologies into a new social synthesis (1974, p. 41). This possibility was never close to realization in the recombinant DNA debate; throughout, by their divided action or by inaction, official bodies of redress facilitated stalemate rather than synthesis.

Turner's less positive notion of reintegration—"the social recognition and legitimatization of irreparable schism between contesting parties"—was not realized either. Instead of reintegration, there was a seemingly permanent clash of purposes, an uneasy truce that left open the question whether this particular conflict had been settled without permanently damaging the invisible bonds of community, the ultimate binding forces of society, that Turner calls *communitas*. To Turner, "the coherence of a completed social drama is itself a function of communitas. An incomplete or irresolvable drama would then manifest the absence of communitas" (1974, pp. 41, 50; see also pp. 46–47).

A quarter of a century ago, E. E. Schattschneider recognized the existence of "unresolvable [political] conflicts" (1975, p. 122); at the same time he was confident that American society would survive these conflicts with unimpaired, perhaps with enhanced, vigor. His confidence stemmed from a belief that steady, deepening, and society-endangering rifts were routinely avoided by constantly shifting interests and mutually canceling stresses. But the recombinant DNA debate may be an instance of such a rift, one of a set of recurring conflicts concerning science and technology, all of which embody a similar clash of purposes: Americans want the benefits of a nearly totally protected science and technology, and none of the risks that nearly total protection entails.

Over the years, the debates over fluoridation, the SST, and nuclear power have been energized less by particular concerns than by a clash of conflicting ideologies more or less common to all: many statements from the nuclear debate can be transposed to the fluoridation debate "simply by changing 'radiation' to 'fluoridation' and 'AEC [Atomic Energy Commission]' to 'PHS [Public Health Service].'"[7] And scientific and technical experts, rather than reassuring the public or helping to resolve these conflicts, seem merely to align themselves in support of one ideological side or another. Thus the forces generated by these debates may be not mutually canceling but cumulative, steadily widening a society-endangering rift, an ideological San Andreas fault beneath a fragile consensus. On this analysis, which side wins a particular debate is immaterial; whoever wins, society is the loser, becoming thereby less cohesive, less viable.

This analysis of the recombinant DNA debate confirms and extends Farrell and Goodnight's analysis of the Three Mile Island incident (1981, pp. 283–284, 287). Both analyses highlight the failure of technical knowledge to resolve social, ethical, or political dilemmas. The Three Mile Island study documents this failure in a bounded event, six days long. Farrell and Goodnight provide a microanalysis of one phase of social drama, the breach. In contrast, my study of the recombinant DNA debate is a macroanalysis of a "complete," though failed, social drama, one that confirms Farrell and Goodnight in a different and wider context. Both analyses conclude that conflicting societal assumptions were responsible for the failed attempt to deal with threatened disaster. For Farrell and Goodnight, these assumptions are embodied in "root metaphors" of industrial expansion, energy needs, and ecological beliefs, root metaphors I would trace to seventeenth-century quarrels between science and religion.[8]

But in their central point, the two analyses coincide: the crucial question is not whether it is in the interest of society to offer to science and technology almost absolute protection; rather, it is whether we can discard the hope that science and technology can solve social problems, whether instead we can find the means to obtain a consensual answer to the questions, other than scientific and technical, that science and technology leave in their wake.

The answer lies, in part, in a deeper examination of the nature of social drama. We do Turner's concept full justice only if we remark that, in cases where science and society clash, the lines blur, the audience participates in the drama. To the extent that we legitimately participate in the conflicts between science and society, we are entitled to initiate actions appropriate to our interests; we are entitled to ask "for a way of controlling and directing research which open[s] up all laboratories to community involvement in their direction; and plan[s] work by a combination of the tripartite structure of decision making by scientists and technicians in the lab itself, by the community in which the lab [is] embedded and by discussions of overall priorities and resources at a national level" (Rose 1987, p. 16).

A willingness to trust such public consensus over the consensus only of experts, no matter how well informed, implies that the democratic process has a legitimate role in the determination of the general direction of scientific work. Resistance to this thesis is understandably common, not only among scientists but among researchers generally. This resistance seems especially justified when, in individual instances—recombination, fluoridation—the science or the technology seems benign, the public outcry largely hysterical. In these instances, we must remind ourselves that the real issue is not the effect of particular initiatives; it is the arrogance of experts, their attempt to circumvent in their own interests the checks and balances of an open society.

Epilogue: Reference without Reality

In *The Concept of Mind* Gilbert Ryle argued that the mind-body problem rested on a misconception, a confusion of categories. By implication, many knotty philosophical problems rested on similar misconceptions: when linguistic analysis was properly applied, these problems would disappear. W. D. Gallie took exception to the generality of this view; he argued that a class of concepts central to our culture would continue successfully to resist Rylean analysis. These concepts were not misconceived: they were, in Gallie's telling phrase, "essentially contested." Appraisive, not descriptive, essentially contested concepts were internally complex and open to various interpretations. But this variety was not the result of confusion: all sides in the ensuing disputes presumed that they were arguing about the same concept, and all traced their claims to the same source or exemplar.

Christian doctrine is Gallie's example of an essentially contested concept. It is clearly appraisive, internally complex, and variously interpreted; yet the many denominations of Christianity concur that they are disagreeing about the same doctrine, derived in each case from the same exemplar. As other instances of essentially contested concepts, Gallie chose "work of art" and "democracy."

To these he might have added the concept of realism, a source of persistent debates in philosophy of science. Do the central terms of science genuinely refer? Do they select just those aspects of the structure of an underlying reality that actually cause the phenomenon in question? Is this reality essentially independent of our perception of it? The variety of answers to these questions testifies to the internal complexity of the concept of realism.

As with all essentially contested concepts, this complexity is not primarily the result of confusion: all disputants agree that they are discussing the same subject, an agreement that stems from a generally shared tradition of philosophical debate. And the concept of realism is certainly appraisive: debates over its significance turn on the question of whether language in one of its functions describes the real; and success in these debates depends only on our estimates of the quality of the arguments for one side or another.

Although the concept of realism is essentially contested, at any one time the arguments for and against realism need not be equally convincing. Indeed, so unconvincing are the traditional defenses of realism that their repair motivates considerable activity in contemporary philosophy of science. But another response is possible: the construction of an intellectually respectable position that does not require the existence of mind-independent objects, a position skeptical, not indeed of knowledge, but of the possibility of metaphysics. This position would be appropriate to a view of science that is thoroughly rhetorical, a view that knowledge is a matter of persuasion and consensus (not *mere* persuasion or *mere* consensus, as some realists would have it).

In this chapter I outline a philosophically defensible position that licenses this radical rhetorical interpretation. By tracing a line from Protagoras to Nietzsche and beyond, a rhetorical epistemology can, doubtless, be responsibly reconstructed. But I will not attempt to vindicate radical rhetorical interpretation by mining traditions alien to realism; rather, I will build my case from within the stronghold of analytical philosophy, constructing a position in which rhetorical interaction is constitutive of knowledge. I make two claims: first, that the philosophical positions of contemporary realists do not entail the realism they espouse; second, that a position consonant with the rhetorical construction of reality may be plausibly derived from their work.

A Critique of Metaphysical Realism

We need look no further than Richard Boyd for an intelligent defense of a robust metaphysical realism. According to Boyd, the success of science is extraordinary and requires an explanation; the best explanation is that science tells a true story about the causal structure of the world. In this most robust version of realism, it is not simply the *terms* of science that refer to an underlying causal reality. The relationships among these terms also refer. Not only do force, mass, and acceleration refer; their relationship refers as well: $F = ma$.

In this most robust version, there can be no genuine discontinuity between the physics of Newton and that of Einstein: $F = ma$ applies both to the conserved mass of Newtonian physics and to the convertible mass of Einsteinian physics. But if Einstein is right, isn't Newton wrong? Are we really justified in saying that the classical relationship between force, mass, and acceleration is reproduced in relativity? We are, if the terms and relationships of science genuinely refer; in that case, we must say that Einstein amended Newton, that he limited the range of application of Newtonian mass to systems with a velocity considerably below that of light.

Against Boyd's position, there are two general arguments:

1. To avoid circularity, metaphysical realists must assume an epistemic access to the causal structure of the world independent of that which science affords. But metaphysical realists do not claim such access; nor could they without resorting to arguments that neither they nor any contemporary philosopher would find acceptable.

2. Metaphysical realism commits its defenders to the following induction: "Time after time, scientific theories genuinely and accurately predict. Either these predictions are simply miraculous, or scientific theories really refer to the causal structure of the world." But this induction rests on a mistaken reading of the history of science. It assumes that science is a generally successful enterprise, a methodological stance that virtually guarantees the production of theories that are largely true. But the history of science is not like that: "overwhelmingly, the results of the conscientious pursuit of scientific inquiry are failures: failed theories, failed hypotheses, failed conjectures, inaccurate measurements, incorrect estimations of parameters, fallacious causal inferences, and so forth." Indeed, "the problem for the realist is how to explain the *occasional success* of a strategy that *usually fails*" (Fine 1984, pp. 89, 104).

Metaphysical realists also misread the history of science when they assume a connection between successful reference and scientific success. In current science, "atoms" genuinely refer, "ether" does not. But the genuineness of his reference did not prevent Dalton from making false claims about atoms, and ether theories—fictions even to realists—were remarkably successful throughout the nineteenth century (Laudan 1984, pp. 221–228).[1]

These historical arguments bear directly on the typical realist claim that science progresses, that later successful theories contain earlier successful theories within their scope. Einstein's mass and Newton's are ontologically incompatible. Nor will it do to equate the m's of $F = ma$ and $e = mc^2$, for this equality in formulas is fully compatible with the claim that in both cases m is a convenient fiction.

Toward a More Defensible Realism

In the face of this assault, contemporary realism has adopted two argumentative strategies. According to the first, some but not all of science refers; according to the second, all of science, indeed all of knowledge, genuinely refers, but the notion of reference must be suitably revised.

Limiting Reference

Ian Hacking and Nancy Cartwright attempt to render realism plausible by drawing a philosophical line: on one side are those scientific terms and relationships that really refer; on the other, those that do not. For Hacking only those scientific entities are real that can be manipulated by means of experiment. A polarizing electron gun demonstrates that parity is violated in a weak neutral current; therefore, electrons are subject to experimental manipulation; therefore, electrons are real. But neutral bosons are not real: "nobody can yet manipulate a bunch of neutral bosons, if there are any" (Hacking 1983, p. 272; see also 1986 and 1987). Cartwright draws an analogous line between scientific fact and fiction. In her view, the terms and relationships of lower-level laws, those relatively close to experience, really refer; moreover, these lower-level laws are causal: "causal laws cannot be done away with, for they are needed to ground the distinction between effective strategies and ineffective ones" (1983, p. 22). But global theories in physics, high-level generalizations like

$F = ma$, are useful fictions, designed only to give order to lower-level laws.

The Hacking-Cartwright line permits us to accept a reasonably accurate account of the history of science without wholly abandoning realism. But the drawing of such a line does not address either of the general arguments against realism, both of which hold as much against Cartwright and Hacking as against Boyd, against a modified as against a robust metaphysical realism. The issue seems moot in any case, since no such line can be drawn: "What I want to know, [says Larry Laudan in a letter to Cartwright] is what *epistemic* difference there is between the evidence we can have for a theoretical law . . . and the evidence we can have for a theoretical entity—such that we are warranted in concluding that, say electrons and protons exist, but that we are not entitled to conclude that theoretical laws are probably true. It seems to me that the two are probably on an equal footing epistemically" (quoted in Cartwright 1983, p. 97). In other words, the question has been begged: you cannot draw the Hacking-Cartwright line unless you have independent access to mind-independent entities.

Revising Reference

To rescue realism, Hilary Putnam and Donald Davidson alter the notion of reference.[2] Putnam categorically denies metaphysical realism: "You can't single out a correspondence between two things by just squeezing *one* of them hard (or doing anything else to just one of them): you cannot single out a correspondence between our concepts and the supposed noumenal objects without access to the noumenal objects" (1981, p. 73). Nevertheless, Putnam insists that "tables and chairs . . . exist just as much as quarks and gravitational fields, and the fact that this pot of water would have boiled if I had put it on the stove and turned on the flame is as much a 'fact' as is the circumstance that the water weighs more than eight ounces" (1987, p. 37). But in virtue of what are facts facts, and counterfactuals facts? In virtue, says Putnam, of internal realism, a realism wholly compatible with "conceptual relativity" (1987, p. 17). To an internal realist, questions concerning objectivity make sense, but only *"within* a theory or description" of the world; "'objects' do not exist independently of conceptual schemes" (1981, pp. 49, 52).

According to Putnam, you cannot have facts without mind-independent entities, and you cannot have mind-independent entities
without conceptual schemes. Very well—but then mind-independent
entities cannot be independently characterized. Putnam describes
truth, the relationship between facts and reality, as *"ultimate goodness
of fit"* (1981, p. 64). But how can this metaphor be cashed out, if facts
are all we can know? What fits what? How well? And how do you
know? Putnam says that "'objects' themselves are as much made as
discovered" (1981, p. 54). But the point is, we can never tell how
much. This is a version of Putnam's own criticism of metaphysical
realism.

There is a second argument against Putnam, one by a fellow realist.
Donald Davidson argues that conceptual relativity cannot have a
coherent interpretation for the realist, for whom there can be "at
most one world" (Davidson 1984 p. 187).[3] Conceptual relativity entails
a dualism of "scheme and content, of organizing system and something waiting to be organized" (p. 189). But if there is only one world,
this is impossible: we can organize objects *within* this world, just as
we can organize the clothes in a closet; but we can no more organize
the world itself than we can organize the closet itself (p. 192). Unlike
Putnam, Davidson understands that "conceptual relativity" entails
"relative truth," a phrase that must be an oxymoron to realists. Philosophy must aim at the only target appropriate to realists: "absolute
truth" (p. 225).

According to Davidson, networks of utterances can accurately
describe the way the world is, the reality beneath appearances. Moreover, it is within such networks that reference is fixed. In other words,
reference is a by-product of the truth of utterances; it is not, as others
have hoped, "*the,* or at least one, place where there is direct contact
between linguistic theory and events, actions, or objects described in
non-linguistic terms" (p. 219). There is no such place.

Not that we lack access to reality through language: "successful
communication proves the existence of a shared, and largely true,
view of the world" (Davidson 1984, p. 201). It follows that "by
studying the most general aspects of language we will be studying
the most general aspects of reality" (p. 201), that a correct theory of
language is also a correct description of reality: "What a theory of
truth does is describe the pattern truth must make among the sentences, without telling us where the pattern falls. So, for example, I

argue that a very large number of our ordinary claims about the world cannot be true unless there are events. But a theory of truth, even if it took the form I propose, would not specify which events exist, nor even that any do" (p. 214).

Davidson's argument rests on a hidden analogy between what we see and what we know to be the case. The coin is circular, despite its oblong look; the stick immersed in water is straight, despite appearances; by analogy, realists say, the world may look this way or that, but these appearances are only manifestations of an unchanging reality, Substance, Extension, Noumena.[4] Davidson is motivated by an analogous metaphysical faith; the most general aspects of language, when we discover them, will accurately reflect the central features of an underlying reality.

But the analogy between commonplace perception and metaphysical speculation does not hold. In the case of the coin and the stick, there is, for realist and nonrealist alike, an independent fact of the matter: measure the coin, grasp the stick. In the case of the world as a whole, there is no independent fact of the matter, no fact separate from that world, no basis for comparison. This is a version of Davidson's own argument against conceptual schemes.

But even if the analogy did hold, it would not establish realism as an appropriate metaphysics for science. Physicists and realist philosophers may share the same "largely true" view of the world, as Davidson implies (1984, p. 201), but so do Tarot readers, phrenologists, and skeptics. Suppose I believe that coins really do change shape when turned, that sticks really do bend when immersed in water: I can still have a predictive physics and a "largely true" picture of the world. But such a physics is rightly obnoxious to both science and common sense.

Why Scientists Are Realists

In his old age, in a volume expressly designed to commemorate his philosophy of physics, Einstein made two seemingly contradictory statements. In his "Autobiographical Notes," he apparently endorses realism: "Physics is an attempt conceptually to grasp reality as it is thought independently of its being observed. In this sense one speaks of 'physical reality'" (1959, p. 81). His "Reply to Criticisms," however, apparently undercuts this realism: "After what has been said, the 'real' in physics is to be taken as a type of program, to which we are, however, not forced to cling *a priori*" (p. 674). When this apparent contradiction between Einstein's statements has been successfully resolved, we will have one aspect of an alternative to contemporary realism, an alternative compatible both with the practice of science and with its rhetorical analysis: we will understand why scientists are realists.

According to Arthur Fine, Einstein's realism has two unwavering central tenets: the natural world is independent of observers, and it is a world wholly subject to determinate rather than probabilistic laws.[5] But these tenets are not part of a credo about the true state of nature; they are instead "a family of constraints" on theories that Einstein would find acceptable. In short, Einstein's realism is motivational; it is a faith in the access of reason to reality: "I have no better expression than the term 'religious' for this trust [*Vertrauen*] in the rational character of reality and in its being accessible, at least to some extent, to human reason. Where this feeling is absent, science degenerates into senseless [*geistlose*] empiricism. Too bad [*Es schert mich einen Teufel*] if the priests make capital out of it. Anyway, there is no cure for that [*ist kein Kraut gewachsen*]."[6]

Motivational realism is the belief that the existence of mind-independent entities is a regulative principle uniquely suitable to the conduct of science. Intent need not be involved. It is not that scientists intend to create realist theories; rather, the possibility of such theories is the psychological anchor that makes a life in science meaningful: "Realism [is] among the prerational springs of human behavior (not, of course, among the irrational ones), those springs that we often conceive of not just as the source of creativity, but also as the source of deep satisfaction in creative endeavors" (Fine 1986, p. 110).

Against the possible generalization of motivational realism to all of science, quantum mechanics may seem a barrier. Certainly, quantum mechanics blocks motivational realism in its full, Einsteinian form. One central tenet must certainly go: there can be no search for general laws of a particular nature—deterministic laws, for instance. But what of the other central tenet, the independence of the observer? This is not undermined by the uncertainty principle, which prohibits only the simultaneous fixing of subatomic position and momentum with an arbitrarily high precision. Nevertheless, how can we contend that quantum physicists are motivated by a faith in an observer-independent reality? By any standards, their deliberate adoption of nonrealism has been all-pervasive and extraordinarily successful.

Appearances are deceptive. Let us take as an example the search for the elusive quark. In 1969 the Dutch physicist J. J Kokkedee spoke in accents typical of quantum nonrealism when he said of the quark model that it "should . . . at least for the moment, not to be taken for more than what it is, namely the tentative and simplisitic expression of an as yet obscure dynamics underlying the hadronic world [of subatomic particles]" (quoted in Pickering 1984, p. 91). In 1969 quarks were a convenient symbol; by 1974 they were nearly real: "There is a great deal of evidence for, and no *experimental* evidence against, the idea that hadrons consist of quarks . . . Let us assume it is true," said Feynman. By 1982, a further step toward realism had been taken: "The quark model gives an excellent description of half the world," said George Zweig (quoted in Pickering 1984, pp. 114, 147; see also p. 270). In less than two decades, then, in a sense at least as robust as that of Hacking and Cartwright, quarks had become real to quantum physicists. Their nonrealist statements notwithstanding, they shared with their contemporary brethren the historical task of revealing the causal structure of the world (see also Fine 1986, pp. 125–126).

Motivational realism explains why scientists believe that their theories must converge on the one real world: it explains Steven Weinberg's search for the "one true theory"; Sheldon Glashow's references to the "apparently correct theory of elementary particle physics"; Stephen Hawking's hope for a "complete, consistent and unified theory of the physical interactions which would describe all possible observations" (quoted in Galison 1983, pp. 46–47). Motivational realism finds the unity of science, of classical and quantum physics, of Aristotelian and evolutionary biology, of alchemy and chemistry, not in the jumble of ontologies that science has variously endorsed, but in the common conviction of scientists that their work has as its goal the discovery of the causal structure of the world expressed in the form of general laws.

Scientific Realism Rhetorically Defined

Because motivational realism does not tell us what makes a scientific theory true, it demands supplementation: we still need an account of scientific truth compatible with thoroughgoing rhetorical analyses. Such an account need not turn its back on the analytic tradition; indeed, the account I offer will bear a strong family resemblance to the views of Quine, Davidson, and Putnam. No paradox is involved: although such views are compatible with metaphysical realism, they do not entail it.

In "Two Dogmas of Empiricism" Quine makes a point that is central to a view of scientific truth compatible with rhetorical analysis: "Total science, mathematical and natural and human, is . . . under-determined by experience. The edge of the system must be kept squared with experience; the rest, with all its elaborate myths or fictions, has as its objective the simplicity of laws" (1961, p. 45). This view is largely shared not only by Davidson (1984, pp. 230–231) but also by Putnam: "'Truth' [is] . . . some sort of ideal coherence of our beliefs with each other and with our experiences *as those experiences are themselves represented in our belief system*" (1981, pp. 49–50).

What sense can we make of edges squared with experience, Quine's metaphor? It certainly implies that meaning contains an irreducibly nonlinguistic component; as Putnam says: *"meanings just aren't in the head"* (1981, p. 19). But when experiences are represented in our belief system, they *are* in our heads, and nowhere else. Surely, it is only in our heads that such representations are possible, and it is only on account of such representations that our experiences can have meaning. To say that our theories must conform to our experiences is not to say, What is in *here* must conform to what is out *there*. It is to say rather that our theoretical utterances must cohere with our observation utterances, those "on which all speakers of the language give the same verdict when given the same concurrent stimulation" (Quine 1969, pp. 86–87).[7] If the verbal network that is science contains the appropriate theoretical and observation utterances in the appropriate configuration, then this network represents the truth of science, *so far as we know*. Within the analytical tradition there exists an intellectual space for the rhetoric of science.

Those who resist the notion that science is fundamentally rhetorical point to the "brute facts": planes fly, men cannot have babies, Snell's Law will not be repealed—surely we cannot be skeptical about such facts. But a rhetoric of science denies none of these. No theory of physics can ignore flight; no theory of biology can turn its back on sex; no optics can dismiss refraction. The claim of rhetoric is that the phrase "brute facts" is an oxymoron. Facts are by nature linguistic—no language, no facts. By definition, a mind-independent reality has no semantic component. It can neither mean, nor be incorporated directly into knowledge. Incorporation by reference is the only possibility: candidate utterances must refer to a mind-independent reality in fact or in principle, a reference earned in a manner approved by relevant epistemic communities, in this case communities of scientists who superintend disciplinary domains. Only such utterances can become part of bodies of knowledge regarded as objective.

Rhetoric of science does not deny that the brute facts are critical to science. Low-level generalizations like Snell's Law, essentially descriptive enterprises like taxonomy and observational astronomy, are part of every science precisely because no science can afford to ignore such generalizations and such enterprises. Because the brute facts are constraints on any theory, they tend to persist, even through scientific revolutions. Whatever the interpenetration of scientific theory and fact, there is a sense in which Ptolemy, Copernicus, Kepler, and Newton explain the same night sky. It is to this sense that I appeal. What is stable in science is not the posited world of physical objects, an ontology that changes as theories change, but precisely the much-denigrated world of appearances, the only world with which science must square itself.

To rhetoricians, science is a coherent network of utterances that has also achieved consensus among practitioners. Putnam is at pains to dismiss the views of those for whom "'true' is just a name for what a bunch of people can agree on" (1987, pp. 17–18). But to say that scientific knowledge represents a consensus concerning the coherence and empirical adequacy of scientific utterances, that the various methods of science are essentially consensus-producing, is not to denigrate science; it is rather to pay tribute to the supreme human achievement that consensus on complex issues represents, an achievement to which this book, I hope, bears witness.

The truths of science, then, are achievements of argument. To this, Putnam nearly agrees: "An argument between an intelligent liberal and an intelligent Marxist will have the same character as a philosophical dispute at the end . . . And we all do have views in religion, or politics, or philosophy, and we all argue them and criticize the arguments of others. Indeed, even in 'science', outside the exact sciences, we have arguments in history, in sociology and in clinical psychology, of exactly this character" (1981, p. 112). Why exclude the exact sciences? Because, unlike their "inexact" counterparts, the exact sciences can "appeal to *public* norms" of rationality, norms that are, presumably, the various methods of science (Putnam 1981, p. 111). But this says only that the exact sciences are the product of a second-order agreement: an agreement concerning agreement. This guarantees only an elevated sense of certainty—not epistemological superiority. By calling some sciences exact, Putnam smuggles into his rhetoric an epistemological superiority he has not earned through argument.

When scientific truth is seen as a consensus concerning the coherence of a range of utterances, rather than the fit between the facts and reality, conceptual change need no longer be justified on the basis of its closer approximation to that reality. It is instead the natural result of the persuasive process that is science, a persistent effort to renew consensus despite a constant influx of potentially disruptive utterances. Because this influx often threatens an existing consensus, one in which men and women have a serious professional investment, any new agreement is an achievement. At the limit of professional investment, we have the cluster of theory absolutely central to research programs; in this case, a new consensus is understandably infrequent.

Still, relatively significant conceptual alterations require no more philosophical or rhetorical apparatus than relatively insignificant ones. Especially to be avoided is the popular notion of competing paradigms, of clashing conceptual schemes: "If we choose to translate some alien sentence rejected by its speakers by a sentence to which we are strongly attached on a community basis, we may be tempted to call this a difference in schemes; if we decide to accommodate the evidence in other ways, it may be more natural to speak of a difference of opinion" (Davidson 1984, p. 197). On this reading, "incompatible" conceptual schemes are merely ways of dramatizing change.

Given this theory of truth, we can, without endorsing realism, legitimately redescribe realist analyses in rhetorical terms. Let us take as an example Hacking's analysis of gravitational lenses. The astrophysicist Edwin Turner is dubious about the ultimate scientific value of these lenses; nevertheless, he does not doubt their existence: they are "a class of phenomena whose very existence depends on rare cosmic accidents," accidents that "occur when two or more objects at different distances from the earth happen to lie along the same line of sight, and so coincide in the sky" (1988, p. 54). Ian Hacking does doubt the existence of these lenses; they cannot be real entities, of whose existence "we are convinced . . . because of a large number of interlocking low level generalizations that enable us to control and create phenomena." For example, it is such generalizations that convince us of the existence of the microscopic entities we manipulate in laboratory experiments (1983, pp. 186–209; see also 1987 and 1986).

Although Hacking's criterion fails in its metaphysical task of singling out the really real, it captures a genuine insight: it tells us what convinces scientists of the reality of the physical objects with which they deal. Interpreted thus, Hacking's realism is fully compatible with a rhetorical perspective; by analogous redescriptions, the analyses of other realists can be rendered compatible also. Redescription gives us a rhetorical interpretation for Putnam's quarks, Davidson's patterns, Cartwright's lower-level laws.[8]

This rhetorical view of science bears a close resemblance to the "radical relativism" of Nelson Goodman's *Ways of Worldmaking* (1978, p. x). In this book Goodman systematically diminishes the importance of truth: "far from being a solemn and severe master, [it] is a docile and obedient servant" (1978, p. 18); "like intelligence, [it] is perhaps just what the tests test" (p. 122). In fact, in science, in cases "where truth is too finicky, too uneven, or does not fit comfortably with other principles," we may choose "the nearest amenable and illuminating lie" (1978, p. 121). Moreover, because truths conflict, "truth cannot be the only consideration in choosing among statements" (p. 120).

Science is less a matter of truth than of making worlds. In the absence of "a ready-made world," waiting to be discovered (Goodman 1978, p. 94), new worlds are necessarily constructed from the old. Making is always a remaking that depends heavily on methods sanctioned by tradition: "Without the organization, the selection of relevant kinds, effected by evolving tradition, there is no rightness or wrongness of categorization, no validity or invalidity of inductive inference, no fair or unfair sampling, and no uniformity or disparity among samples. Thus justifying . . . tests for rightness [a more general term that includes truth] may consist primarily in showing not that they are reliable but that they are authoritative" (pp. 138–139).

Goodman shows us how we can stay within the analytical tradition
without privileging either truth or scientific knowledge, and without
accepting realism in *any* of its forms.

Traditionally, in the hegemony of knowledge, science has been the
master, dialectic and rhetoric the servants; traditionally, it is in terms
of the scientific syllogism that its dialectical and rhetorical counterparts have been defined. From this perspective, a resort to dialectic
is an intellectual step down: the premises of dialectical syllogisms may
be less than true, and mutual agreement through reciprocal understanding is insufficient for truth. A resort to rhetoric represents a
further descent: rhetorical syllogisms may lack either of their premises or their conclusions; this defect leads to their radical dependence
on the beliefs of individual audiences, a condition guaranteed to be
at odds with truth. Moreover, rhetoric so construed emphasizes
agreement to the detriment of reciprocal understanding. Finally,
rhetorical effectiveness relies importantly on the character of the
speaker and the emotional disposition of the audience—rightly matters of unconcern to both the scientist and the dialectician.

When rhetoric is so defined, it is little wonder that the traditional
view of Aristotle's masterpiece prevails: "a curious jumble of literary
criticism with second-rate logic, ethics, politics, and jurisprudence,
mixed by the cunning of one who knows well how the weaknesses of
the human heart are to be played upon."[9]

But suppose we alter the judgment of tradition; suppose, instead,
we define dialectic and logic in terms of rhetoric. From this perspective, dialectic and logic are rhetorics designed for special purposes:
dialectic, to generate the first principles of the special sciences; logic,
to derive from these principles true statements about the causal structure of the world. When logic and dialectic are defined in this way,
rhetoric cannot be dismissed as defective. On the contrary, it becomes
the more general term that includes logic and dialectic, defined now
as rhetorics for special purposes.

It is this revision of the traditional relationship between logic,
dialectic, and rhetoric that I endorse in this book. My rhetorical
analyses show how the sciences construct their specialized rhetorics
from a common heritage of persuasion. By means of these, the

sciences create bodies of knowledge so persuasive as to seem unrhetorical—to seem, simply, the way the world is. But however much scientists require the justification of realism, rhetoricians are realists only at their peril: for them, realism must remain an analytical target, a rhetorical construct like any other.

Notes

1. Rhetorical Analysis

1. Of all workers, Bruno Latour makes the rhetorical orientation of his studies most explicit. In *Science in Action* Latour places science among a web of activities that includes virtually every center of influence in human affairs. Persuasion, which constitutes each center, is also the binding force within the web.

2. In *The Social Basis of Scientific Discoveries*, Augustine Brannigan makes the analogous point that scientific discoveries are social constructions based on the novelty, validity, and plausibility of candidate objects or events in the context of recognized programs of research.

3. A broader definition of *ethos* is usual, one that includes matters of value; however, for expository convenience, I categorize such matters under *pathos* later in this section. Nothing significant rides on this idiosyncratic allocation.

2. Analogy in Science

1. On early developments in the use of analogy, see Lloyd 1971. On Aristotle's use of analogy in his scientific treatises, see McKeon 1949; on analogy in modern science, see Hesse 1966; on the uses of analogy in non-scientific discourse, see Perelman and Olbrechts-Tyteca 1971, pp. 371–398.

2. This is the received view, as given in Perelman and Olbrechts-Tyteca 1971, p. 396. In *Models and Analogies in Science* (1966), Hesse makes a perceptive case for the incorporation of analogy into theory. In Chapter 5 I shall endorse a view of analogy at least as radical as hers.

3. For some cautions about the excessive laudation of quantification, see Douglas 1971 for the social sciences; for the natural sciences, see Kuhn 1977, pp. 178–224.

3. Taxonomic Language

1. Although scientific characterization of the species is a subject of serious controversy (see Hull 1981a, 1981b, 1983a, 1983b, 1984), I think these views are defensible.

2. This orthodox view is not held by radical pheneticists (see Hull 1981a).

3. Systematists not only use characters; they also "[inquire] into the origin and nature of the units with which [they work]" (Mayr 1982, p. 9).

4. According to Hull (1983a), it is a general practice in much of biology to deal with species as natural kinds: "Though comparative anatomists clearly acknowledge that species evolve, they insist that they can go about their business as if they did not" (p. 76).

5. For the application of this view to another group of scientists, see Mulkay and Gilbert 1981, p. 403.

6. I dwell on Popper because scientists do; I am far from holding up his ideas as an accurate way to reconstruct good scientific practice. Whatever the status of evolutionary theory in relation to taxonomy, taxonomists have successfully practiced for centuries a science that is simply not about bold conjectures and crucial observations. Indeed, it is even questionable whether physics, Popper's obvious model, is such a science. Let us glance at one famous instance: the bending of light in a gravitational field, designed as a crucial test of general relativity. In this instance, the data that physicists found so convincing exhibited a scatter far wider than any ordinary canons of proof would allow (Bernstein 1985, pp. 141–146; compare Franklin 1986, pp. 226–243). In the end, as Bernstein makes clear, physicists were persuaded as much by the elegance of the theory as by its alleged resistance to falsification, a resistance that, under different circumstances, could easily have been given the opposite interpretation. What is true in physics is true *a fortiori* in evolutionary taxonomy: "The reason for taxonomists paying so much attention to the works of Popper is that they think that they can use his Principle of Falsifiability to show that *their* classifications are truly scientific, while those of their opponents are not" (Hull 1981a, p. 142).

7. These two meanings of rhetoric correspond to Husserl's distinction between two sorts of reflection: the natural and the radical (Carr 1974, pp. 16–27).

4. The Tale of DNA

1. In *The Rhetoric* Aristotle equates *energeia* with *prosopopoeia* (1975, pp. 406–407). By the time of Quintilian, the concept had been generalized to cover any extraordinary vividness in the presentation (1920–1922, II, 435–439).

2. Sayre 1975, p. 190; but see pp. 129, 145–146, 162–163; for other possible discrepancies in detail, see Olby, 1974, pp. 346, 350–351, 354, 389, 411–412; see also P. Pauling 1973, Perutz 1967, and Chargaff 1974.

3. There is some question as to how near Franklin was to the solution.

Although she knew the B form was helical, she did not realize that the helices ran in opposite directions or that the bases aligned themselves according to Chargaff's ratios. See Portugal 1977, p. 265; see also Olby 1974, p. 351.

4. According to Branningan's sociological model, Pauling is constructing the Watson-Crick DNA structure as a discovery by perceiving and expressing its novelty, validity, and plausibility within a recognized research program.

5. Because the Grimm tales are literary versions of folk materials, they ought not to be confused with the folk materials themselves. See Ellis 1985, p. 70, for a summary of the Grimms' relation to their sources.

6. In *Rosalind Franklin*, Sayre says that the writing of the epilogue was "pressed upon" Watson (1975, p. 194). However, her transcript makes Watson's voluntary agreement clear (pp. 218–219).

7. Equally, those who were appalled, like Chargaff and Sinsheimer, were reacting to a view against which they had very strong feelings indeed.

8. For another "masterpiece of understatement," this time in physics, see the abstract of Lee and Yang's Nobel Prize–winning paper on the non-conservation of parity in weak interactions, quoted and commented on in Franklin, 1986, pp. 14–15.

9. In Pinch's terminology (1985b), the externality of the observation increases, increasing both its profundity and the risk of challenge. In the case of the double helical structure, however, the increase in profundity cost nothing: all challenges must be to the structure itself.

5. Style in Biological Prose

1. These features are assumed to be uniform in all criticisms of scientific prose, and are part of the conclusions of all studies of that prose of which I am aware. For critiques, see Bram 1978, David 1976, King 1978; for studies, see Quirk et al. 1979, pp. 807–808, 933–934, Kinneavy 1971, Lin 1979, Wright 1985; for suggestions for improvement, see Day 1979, Williams 1985.

2. Strawson 1974, pp. 105–109; see also Strawson 1971. The application of Strawson's theory to scientific prose is not Strawson's but my own.

3. This is, of course, a rational reconstruction of a complex social and psychological process.

4. From the files of one issue of a leading biological journal. At the price of complete anonymity, the editors kindly permitted me to examine their files. To disguise identification, I have substituted biological nonsense for the biological sense of the quoted passages. For an earlier analysis of this material, see Gross 1984.

5. For some insight into the passive voice, see Lyons 1978, Johnson-Laird 1968, and Sinha 1974.

6. The interpretations in this section owe a considerable debt both to traditional and to more radical commentary. For traditional commentary, see Bloomfield 1970, Ehrenberg 1977, Gross 1983b, Mahon 1977, Waller 1979, Wright 1977, and, most important, Tufte 1983; for more radical commen-

tary, see Bastide 1985, Ivins 1938, Latour 1986, and Lynch 1985a, 1985b. A useful anthology of more radical views is Latour and de Noblet 1985.

7. The phrase, but not its application to science, appears in Strawson 1974, p. 82.

8. To follow the research, in addition to the papers by Spector and Racker (1980, 1981), see the paper of Rephaeli (1981). To follow the controversy, see Broad and Wade 1982; "Oncogenes" 1981; Racker 1981 and 1983; Vogt et al. 1981; "Inadmissible Evidence" 1981; Kolata 1981.

6. The Arrangement of the Scientific Paper

1. Latour and Woolgar 1979, p. 252. Norms for the arrangement of scientific papers are now typically set by scientific societies: see *Handbook* 1978; *Style Manual* 1964; *General Notes* 1950, 1965. Day 1979 is a good book on the subject.

2. Knorr-Cetina 1981, p. 118; see also Bazerman 1981, Gusfield 1976, Medawar 1964, Woolgar 1981.

3. In Chapter 3 I show that the form of the descriptive paper closely imitates that of the experimental report.

4. Boyle 1965, pp. 336–342. Boyle's piece is not a paper but part of a book. However, I am concerned here not with the historical accidents of publication or the presence or absence of headings, but with essential form. In the journal *Science,* for instance, papers generally lack headings; nevertheless authors are advised to "provide a brief outline of the main point of your report in a short introductory section, then describe your experiments and the results and conclude with a discussion" ("Instructions for Contributors" 1983, p. xii).

5. Since Bacon, of course, a great deal has happened in philosophy of science, and it may seem obtuse of me to ignore this; but my point is that scientists are equally obtuse, and for good reason.

6. Although the sentiment is Baconian, the phrase is from Leibniz (1976, p. 465).

7. Although the sentiment is Baconian, the phrase is from Boyle (1965, p. 277).

8. For Bacon's notion of arrangement, see Bacon 1937, pp. 371–373; see also pp. 488–489. See also Jardine 1974, p. 174. In the *Proceedings* of the Royal Society, headings are only occasional throughout the nineteenth century; not until 1935 are they regularized, and not until 1950 is the regularization formalized in *General Notes.*

9. Knorr-Cetina 1981, p. 110. Citation and acknowledgments also further this end. In Boyle these are casually displayed, in Nirenberg and Matthaei formally displayed. Bacon implicitly acknowledged the need for citation (1964, p. 126) but feared the adverse effects of the weight of authority (1960, pp. 280–281 and 1964, pp. 126–127). The literature on networks of citations as symbols of research programs is extensive. See, for example, Gilbert 1976 and Small 1978.

10. These are called maxims by Grice, who formulated them; summarized in Lyons 1978, II, 592 ff.

11. See Lyons 1978, II, 592 ff.

12. "*A* associated with *B* by nature, but selected as an index of *B* by human choice ('smoke is an index of fire')" (Leach 1982, p. 12).

13. For a comment on the Abstract, see Woolgar 1981, p. 261; for a seventeenth-century comment on the movement from laboratory events to scientific facts, see Leibniz 1976, pp. 88–89; for a twentieth-century comment, see Latour and Woolgar 1979, pp. 75–86.

14. Three papers in this set do not offer any predictions. In "Do Gravitational Fields Play an Essential Part in the Structure of the Elementary Particles of Matter?" Einstein ends by recording a shortcoming of the general theory (1952, p. 198); in "Cosmological Considerations on the General Theory of Relativity" (1952, pp. 187–188), he suggests a curved, closed universe compatible with relativity theory; in "Hamilton's Principle and the General Theory of Relativity" (1952, pp. 165–173), he concerns himself not with the truth of the theory, but with the elegance of its derivation.

15. Einstein 1954, p. 276. These views are the source of Einstein's opposition to the reigning probabilistic ontology in quantum mechanics; see also pp. 315–316.

16. "Ich überzeugt bin, dass sie im Rahmen der Anwendbarkheit ihrer Grundbegriffe niemals umgestossen werden wird" (Einstein 1959, I, 32–33).

17. Einstein 1954, p. 262; see also p. 227. In the reigning research paradigm, quantum mechanics, Einstein's search for determinate laws has been abandoned. But the abandonment of this search does not affect the arrangement of the theoretical paper. An examination of some papers on renormalization, a quantum mechanical procedure that as much as any typifies the ontological ingenuity at the heart of that science, reveals close formal similarities. The difference, of course, is that the fundamental laws of quantum mechanics are statistical, not determinate (Gell-Mann and Low 1954; Wilson 1971).

18. From 1919 to 1952 there were twelve attempts to measure the bending of light. Because no star measured was less than two solar radii from the surface of the sun, the results are even less certain than my discussion implies (Sciama 1959, pp. 70–71).

19. Laymon 1984, pp. 114, 117. From his exemplar, Laymon draws conclusions very different from mine.

20. Lévi-Strauss 1963, p. 229; see also 1976, pp. 146–197, and Leach 1980, pp. 57–91.

21. See also Feyerabend 1970. Schuster makes a related point about all general methods in science: they are all mythic speech in the sense that none can be true, though all are based on evidence incontrovertible to believers.

7. Copernicus and Revolutionary Model Building.

1. This is also Chalmers's view of the *Dialogue* (1986, pp. 21–23). But Chalmers is far from agreeing with Feyerabend that the *Dialogue* truly rep-

resents Galileo as a scientist. In works like *The Two New Sciences,* Chalmers believes, Galileo used sound reasoning and eschewed propaganda.

2. My view of the Copernican revolution departs not only from Feyerabend but from Stephen Toulmin, in his influential *Human Understanding.* Toulmin asserts that between claims as different as the Ptolemaic and the Copernican, a rational mediation is possible; in fact, in that particular case, such a mediation actually occurred: "If the men of the sixteenth and seventeenth centuries changed their minds about the structure of the planetary system, they were not forced, motivated, or cajoled into doing so; they were given reasons. In a word, they did not have to be converted to Copernican astronomy; the arguments were there to convince them" (1977, p. 105). I do not deny that they were given reasons—more and more reasons, as time went on. But in my view, "forced, motivated, or cajoled" and "given reasons" are equally means of persuasion.

3. Willard 1983, p. 91; but see Rowland 1982, Wenzel 1982, and Zarefsky 1982.

4. The first set of page numbers is to the English translation, the second to the best edition of the Latin text.

5. "quod tanto et tam mirabili consensu perficiatur!" (Hugonnard-Roche 1982, p. 47).

6. "in natura necessariis satisfieri opportunum fuit" (ibid., p. 68).

7. "undique causus appparientium elucentibus . . . nullis aliis assumptis hypothesibus commodius ac rectius demonstraverit" (ibid., p. 64).

8. "quin simul totum systema, ut consentaneum erat, de novo in debitas rationes restitueretur" (ibid., p. 57).

9. "quid a se in his demonstratum sit, et quid tanquam principium sine demonstratione assumptum" (ibid., p. 58).

10. "nisi magnis de causis ac rebus ipsis efflagitantibus" (ibid., p. 81).

11. "tabulas cum diligentibus canonibus sine demonstrationibus" (ibid., p. 85).

12. For a parallel case of myth-making in chemistry, see Bensaude-Vincent (1983) on Lavoisier.

13. "necesse fuit, ut D[ominus] Praeceptor meus novas hypotheses excogitaret" (Hugonnard-Roche 1982, p. 53).

14. "tota mundi fabrica totaque siderum chorea explicata sit" (Prowe 1967, II, 202).

15. "in natura necessariis satisfieri opportunum fuit" (Hugonnard-Roche 1982, p. 68).

16. Compare Chalmers (1986, p. 10) for a similar view.

8. Newton's Rhetorical Conversion

1. In general, the reader can discern my historical biases by following my notes. I should like to mention here some works that escaped quotation but were crucial in forming my views: Cohen 1966; Koyré 1968; Crombie 1961; Kuhn 1977; Lindberg 1976; Descartes 1979; Wallace 1959.

2. Descartes, *Optics,* in Descartes 1965, p. 70. This passage is typical of Descartes's expository technique, and a good example of rhetorical transparency: the analogy is designed to make the science clearer.

3. Throughout, but especially in Descartes 1983/84, pp. 286–288.

4. In French, *expérience* means both "experience" and "experiment."

5. For example, compare Descartes 1965, p. 268, with a letter to Father Mersenne, dated by inference March 1, 1638, in Descartes 1898, II, 29.

6. 1959, I, 169 (a paraphrase of the Latin); see also 1978, p. 93 and p. 506n.

7. 1978, p. 57. It is a delicious irony that in this passage Newton uses Aristotelian terminology against itself.

8. On the rewriting of the history of science to legitimate change, see Graham, Lepenies, and Weingart 1983, especially the papers of Bensaude-Vincent, Galison and R. Laudan.

9. The number of Queries rose from a mere sixteen in the first English edition (1704), to twenty-three in the first Latin edition two years later, to the full thirty-one in the second English edition (1718). See Newton 1979, p. xxxi, and Westfall 1984, p. 641.

10. Quirk, Greenbaum, Leech, Svartvik 1979, p. 401. Newton seems to have borrowed this device from Hooke. See Hooke 1938, pp. 233–240.

11. Newton did not publish his *Opticks* until 1704, a year after Hooke, the last living critic of his early paper, died. A neurotic fear of rejection probably fueled this delay and, in large part, drove Newton's rhetorical ingenuity. Regardless of Newton's conjectured motives, however, the *Opticks* remains a rhetorical masterpiece. For a psychological interpretation of Newton, see Manuel 1979.

12. The words are Schuster's (1986, p. 80); see Feyerabend 1970.

9. Peer Review and Scientific Knowledge

1. For the nature of a regulative speech act, see Habermas 1979, p. 64; for the conditions pertaining to a request, see Searle 1969, p. 66.

2. In compliance with the wishes of the editors of this journal, I have disguised these selections by the judicious omission of identifying nouns and adjectives.

3. Cole and Cole 1981, p. 56. See also Cole, Rubin, and Cole 1978. This view appears in print first in two letters by Ward and Goudsmit (1967, p. 12). For a recent perspective, see Zimmerman 1982, pp. 46–48.

4. It is on this point that journals in the humanities deviate most; in the case of split decisions they are inclined to reject, a decision rule undoubtedly necessitated by the imperative to reject most work.

5. Thomas McCarthy, private correspondence, November 23, 1987.

6. Bach and Harnish 1979, p. 46. Habermas would, of course, use the terminology of Austin and Searle; he would call these constatives. But the refinements of Bach and Harnish embody genuine insights, ones with which

I fancy Habermas would be comfortable. However, nothing in my argument depends on this refinement.

7. Wimsatt traces the difficulty of reconstructing scientific heuristics to this "practice of not describing chains of hypothetical reasoning or discovery in scientific papers" (1980, p. 235).

10. The Origin of *The Origin*

1. This chapter concerns the primary records of a scientific discovery; for a rhetorical analysis of two reconstructions of that process, one autobiographical, one scientific, see Chapter 4. For an apt sociological analysis of discovery, see Brannigan 1981.

2. I do not mean to imply that Harré and Peirce endorse a rhetorical analysis of coming to believe; only that the appropriateness of such an analysis is a legitimate implication of their thought.

3. In transcribed passages, I have marked Darwin's deletions "⟨ ⟩," his additions "{ }," editorial additions "[ ]," end of entry or page "|."

4. Arguably, it is a mistake to apply the term "style" to linguistic strings whose verbal units are relatively free from conscious manipulation. This freedom is certainly conceivable in Darwin's most primitive entries, the ones most resistant to interpretation. Concerning these, punctilious readers might want to reserve a new term, style$_0$. They ought to be reminded, however, that such fastidiousness concerns only the legitimate extension of terms. My argument assumes the growth of conscious control, but at no place does it depend on determining the exact point at which such control begins.

5. 1987a, pp. 69–70; 1987b, pp. 406–407. In these entries, Darwin refers to his discovery of a petrified forest in the Andes as unmistakable evidence that this mountain fastness had once been a coastline.

6. My treatment follows Habermas 1979, pp. 1–68. The validity claim of truthfulness—that I mean what I say—does not apply in the case of a system of internal representations. When we say people lie to themselves, we are employing a trope; you cannot believe a falsehood you tell yourself.

7. In the Notebooks, this development from cryptic to telegraphic is relatively rapid in the case of geology, relatively slow in the case of evolutionary theory: not until 1844 had Darwin produced a theory of evolution so well articulated that he was willing to risk its publication. In contrast, the geological passages in the *Red Notebook* move quickly toward the relative stability of the telegraphic style, a speed consonant with their proximity to publication. For confirmation of this bifurcation in Darwin's intellectual development, see Sulloway 1985.

8. The third leitmotif concerns continuities of behavior among organic beings as expressions of their evolutionary relationship. This is the chief subject of the M and N Notebooks and, most notably, of *Expression of Emotion in Man and Animals*.

9. 1987a, pp. 60–64. The cryptic nature of the entries makes the borders

between them somewhat unclear; however, nothing in my argument depends on exact demarcation.

10. 1987a, pp. 60–61. See note 3 for an explanation of editorial symbols. For some elucidation of the entry on antarctic vegetation, see 1987b, pp. 274–275 and 1985, II, 411–412; on hydrophobia, see 1987b, p. 436; on drought, see 1987b, pp. 155–158.

11. In the Notebooks, and elsewhere before complete formulation, Darwin uses the term indifferently for any statement with theoretical content.

12. These stages are not subject to absolute chronological separation: Darwin's early letters and notes are full of speculation (Sulloway 1985), his Notebooks show a persistent concern for public presentation; and it was not until well after 1842 that he came to understand a crucial aspect of evolutionary theory, the link between species divergence and niche availability (1958, pp. 120–121).

13. Different versions of this view are held by Ghiselin (1984), Gruber (1981), Kohn (1980), and Richards (1987).

14. The boldface means that this is an annotation Darwin added later.

15. Ghiselin's analysis is similarly flawed (1984, pp. 56–57). In fairness, Gruber later supports the essential ambiguity of the Notebooks (1985, pp. 17–18).

16. Kohn 1980, pp. 100–101. The definition is borrowed from Kenneth Shaffner.

17. Subsequently identified as a rhea.

18. For species identifications, see Darwin 1987b, p. 353; corrected in 1962, p. 290.

19. I do not mean to imply that the analysis of ordinary problem solving is less than a complex intellectual challenge (see, for instance, Gruber 1981, 1985).

20. Manier 1978, pp. 157–158. Manier attributes this state to Darwin only at a crucial point in the composition of the Notebooks; it is I who generalize.

21. Swinburne 1984, p. 64. Swinburne is a dualist, but neither this claim nor the arguments that support it depend on this dualism. See Johnstone for a vivid description of the productive conflict between argument and self-maintenance (1978, pp. 107–111).

22. Campbell (1975, pp. 377–378) points this out, but he attributes it to Darwin's personal rhetoric; he does not recognize this general belief as a necessary condition for the fixation of any particular belief.

23. Brannigan's analysis of genius as a folk-explanation of discovery is apposite (1981, pp. 153–162).

11. The Emergence of a Social Norm

1. Geertz 1973, pp. 207, 212–213; see also Collins 1975, Gieryn 1983, Gilbert and Mulkay 1984, Gusfield 1976, Woolgar 1981.

2. In the myth, Bensalem avoids imperialist confrontation by keeping its existence a secret. It has few visitors, and its collectors of scientific informa-

tion, its Merchants of Light, "sail into foreign countries under the names of other nations" (1937, p. 469).

3. The phrase is from Barnfield (1968, p. 53). See Black for some insight into the terminology problem (1962, p. 47).

4. Max Black's theory, which I find convincing in this case, makes metaphoric meaning the product of the interaction of tenor and vehicle. This theory permits the Baconian use of metaphors from trade and imperialism to prepare the way for other uses, very different in meaning.

5. Oldenburg 1965–1973, III, 535–538; IV, 419–424; see also VII, 259–260 and 336–338.

6. Westfall 1984, pp. 446–452; Oldenburg, 1965–1973, II, xxii; V, 374–375; X, xxiv, xxvi, 73; Birch 1968, IV, 58, 84–86.

7. This is an English summary. The full text appears in Huygens 1897, VII, 305–308.

8. Oldenburg, 1965–1973, III, 373; X, 41–43, 282; see also IX, 377–378, and Newton 1959–1977, I, 73; IV, 100.

9. To this day, no such body exists. A typical contemporary effect of its absence is the continuing priority dispute between American and Soviet scientists concerning the discovery of two transuranium elements.

10. To follow the arguments in the "Account," a reader needed a knowledge of both Latin and mathematics; in that sense, Newton's audience was very specialized, and not particularily English. But my point here is that the rhetoric of the "Account" creates its own audience, a fiction of Englishmen patriotic to the point of chauvinism.

11. For another example of the connection between science and chauvinism, the case of the nineteenth-century rivalry between Germany and France, see Bensaude-Vincent 1983, pp. 64–67.

12. This is not to deny that earlier scientists might have seen some personal gain in asserting temporal priority for their discoveries; Galileo is a well-known example (1957, pp. 232–233, 245).

13. For the idea of a revolution in law, I am indebted most to Tigar and Levy (1977). Their account is avowedly Marxist, but it is not seriously distorted, as reference to Holdsworth (1966) and Jenks (1949) makes clear. For the idea of a right, I am indebted to Hart (1980).

14. This exclusivity is also in line with the English practice of primogeniture, whereby the eldest son inherits all. This is far from the case on the Continent (see Pollock and Maitland 1968; Knappen 1964; Smith 1928). In Newton's particular case, such exclusivity may also have had a psychological motive: Newton was a posthumous only son abandoned by his mother and stepfather (Westfall 1984; Manual 1979).

15. Oldenburg 1965–1973, II, 329; III, 537; IV, 422; V, 104, 178; X, 2, 67.

16. Sprat 1667, p. 311; Birch 1968, III, 514; IV, 60, 464; Stimson 1948, pp. 66–67.

17. Birch 1968, IV, 452; 486, 488, 527, 556; 492; 499; 550.

18. Given the state of society and of science at the time, the occurrence

of particular scientific events may be virtually inevitable. Whether these events are called discoveries, simultaneous or otherwise, is, of course, a matter of social attribution. Indeed, unless discovery is a question of social attribution, my analysis makes no sense: the ownership of private events cannot be a matter of dispute.

12. The Social Drama of Recombinant DNA

1. Victor Turner 1974. It should be understood that this analysis extracts Turner's concept of social drama from a set of equally fruitful allied concepts, such as structure, antistructure, and liminality. For another study using Turner, see Frank 1981.

2. Cited in Jones 1961, pp. 187–188 and 195; see also under "time" and "tide" in *The Oxford Dictionary of English Proverbs*.

3. The first phrase is Merton's (1973, p. 25); the sentiment, of course, is not his. The second phrase is from Farrell and Goodnight 1981, p. 295.

4. Marlowe 1910, p. 194; M. Shelley 1963, pp. 18, 46; Hawthorne 1964, p. 344.

5. Watson and Tooze 1981, pp. 58–59, 104, 237; National Academy of Sciences 1977, pp. 82, 249.

6. Watson and Tooze 1981, p. 508. See also Delgado and Millen 1978. The reference to Canute is from a case not about recombination but about more conventional genetic manipulation.

7. Mazur 1973, p. 248; see also Clark 1974; Bytwerk 1979; Green 1961; Mazuzan 1982.

8. See Jones 1961 throughout, but especially pp. 237 ff. Casaubon's views on "the law of man as opposed to the law of the thing" (p. 243) seem especially pertinent.

Epilogue

1. For a pair of papers in which the arguments against metaphysical realism are set forth eloquently and in detail, see Fine 1984 and Laudan 1984.

2. Or, rather, of satisfaction, a term in which predicates and indexicals are also included.

3. Davidson also seems to believe that conceptual relativity is incoherent for anybody; but that is another matter.

4. Although Kant disclaimed knowledge of noumena, he claimed that their existence was intelligible (Scruton 1982, pp. 42–46).

5. In addition, this realism has two secondary features: physical reality must constitute a system in space-time in which distant real entities cannot influence one another causally, and must exclude ontologically disparate entities such as point particles and continuous fields.

6. Quoted in Fine 1986, p. 110. Fine's translation does not convey the colloquial flavor of the German original. For *geistlose* read "mindless," even

"stupid"; for *Es schert mich ein Teufel*, "I don't give a damn"; for *ist kein Kraut gewachsen,* "it's a disease without a cure."

7. This argument does not depend on whether truth is a property of utterances, as I believe with Davidson, or of propositions or sentences. Neither does it depend on a sharp distinction between theoretical and observation utterances, a distinction no longer tenable.

8. Of course, none of this need prevent realists from noticing my omission of what is, from their point of view, the central constraint on agreement in the sciences: the recalcitrance of the physical objects of the only world that is. These issues, one must remember, remain essentially contested.

9. Ross 1971, p. 275. This is not Ross's view. Ross believes that the *Rhetoric* "is not a theoretical work . . . it is a manual for the speaker" (p. 276). Needless to say, I disagree with him on this point.

References

Anscombe, G. E. M. 1957. *Intention,* 2nd ed. Ithaca, N.Y.: Cornell University Press.

Aristotle. 1975. *Aristotle's Posterior Analytics,* trans. J. Barnes. Oxford: Clarendon Press, 1975.

———. 1926. *"Art" of Rhetoric,* trans. John Henry Freese. Reprint, Cambridge, Mass: Harvard University Press, 1975.

———. 1968; 1960a [1934; 1929]. *The Physics,* ed. P. H. Wicksteed and F. M. Cornford, 2 vols. Cambridge, Mass: Harvard University Press.

———. 1960b. *Topica,* ed. E. S. Forster. Cambridge, Mass: Harvard University Press.

Armitage, A. 1962 [1957]. *Copernicus: The Founder of Modern Astronomy.* New York: A. S. Barnes.

Bach, Kent, and Robert M. Harnish. 1979. *Linguistic Communication and Speech Acts.* Cambridge, Mass: MIT Press.

Bacon, Francis. 1962 [1915]. *The Advancement of Learning,* ed. G. W. Kitchin. London: Dent.

———. 1937 [1627]. *Essays, Advancement of Learning, New Atlantis, and Other Pieces,* ed. Richard Foster Jones. New York: Odyssey Press.

———. 1960. *The New Organon and Related Writings,* ed. Fulton H. Anderson. New York: The Liberal Arts Press.

———. 1964. *The Philosophy of Francis Bacon: An Essay on Its Development from 1603 to 1609 with New Translations of Fundamental Texts,* ed. Benjamin Farrington. Chicago: University of Chicago Press.

Bambrough, R. 1966 [1960–1961]. "Universals and Family Resemblances." In *Wittgenstein: The Philosophical Investigations, A Collection of Critical Essays,* ed. G. Pitcher. Notre Dame, Ind: University of Notre Dame Press, pp. 186–204.

Barnfield, Owen. 1968. "Poetic Diction and Legal Fiction." In *The Importance of Language,* ed. Max Black. Ithaca, N.Y.: Cornell University Press, pp. 51–71.

Barthes, Roland. 1968 [1964]. *Elements of Semiology,* trans. Annette Lavers and Colin Smith. New York: Hill and Wang.

———. 1970. "Science versus Literature." In *Introduction to Structuralism,* ed. Michael Lane. New York: Basic Books, pp. 410–416.

———. 1974 [1970]. *S/Z: An Essay,* trans. Richard Miller. New York: Hill and Wang.

Bastide, Françoise. 1985. "Iconographie des Textes Scientifiques: Principes d'Analyse." In "Les 'Vues' de L'Espirit," ed. Bruno Latour and Jocelyn de Noblet, *Culture Technique* 14: 132–151.

Bazerman, Charles. 1988. *Shaping Written Knowledge: The Genre and Activity of the Experimental Article in Science.* Madison: University of Wisconsin Press.

———. 1981. "What Written Knowledge Does: Three Examples of Academic Discourse." *Philosophy of the Social Sciences* 11: 361–387.

Beer, Gillian. 1985a [1983]. *Darwin's Plots: Evolutionary Narrative in Darwin, George Eliot, and Nineteenth-Century Fiction.* London: Ark.

———. 1985b. "Darwin's Reading and the Fictions of Development." In *The Darwinian Heritage,* ed. David Kohn. Princeton, N.J.: Princeton University Press, pp. 543–588.

Bensaude-Vincent, Bernadette. 1983. "A Founder Myth in the History of Sciences? The Lavoisier Case." In *Functions and Uses of Disciplinary Histories,* ed. Loren Graham, Wolf Lepenies, and Peter Weingart. Dordrecht: D. Reidel, pp. 53–78.

Berger, Peter L., and Thomas Luckmann. 1967. *The Social Construction of Reality: A Treatise in the Sociology of Knowledge.* New York: Doubleday.

Berlin, Isaiah. 1978. *Concepts and Categories: Philosophical Essays.* New York: Viking Press.

Bernstein, Jeremy. 1968. "Confessions of a Biochemist," review of *The Double Helix* by James Watson. *New Yorker* 44 (April 13, 1968): 172–182.

———. 1985 [1973]. *Einstein.* Hammondsworth, England: Penguin.

Berry, A. 1961 [1898]. *A Short History of Astronomy from the Earliest Times through the Nineteenth Century.* New York: Dover.

Bettelheim, Bruno. 1977. *The Uses of Enchantment: The Meaning and Importance of Fairy-Tales.* New York: Knopf.

Birch, Thomas. 1968. *The History of the Royal Society for Improving of Knowledge from Its First Rise,* ed. A. Rupert Hall and Marie Boas Hall, A Facsimile of the London Edition of 1756–57, 4 vols. New York: Johnson Reprint.

Black, Max. 1962. *Models and Metaphors: Studies in Language and Philosophy.* Ithaca, N.Y.: Cornell University Press.

Blackburn, Simon. 1984. *Spreading the Word: Groundings in the Philosophy of Language.* Oxford: Clarendon Press.

Bloomfield, L. 1970. *A Leonard Bloomfield Anthology,* ed. C. F. Hockett. Bloomington: Indiana University Press.

Boyd, Richard. 1984. "The Current Status of Scientific Realism." In *Scientific Realism,* ed. Jarrett Leplin. Berkeley: University of California Press, pp. 41–82.

———. 1979. "Metaphor and Theory Change: What Is 'Metaphor' a Metaphor For?" In *Metaphor and Thought,* ed. A. Ortony. Cambridge: Cambridge University Press, pp. 356–408.

Boyle, Robert. 1965. *Robert Boyle on Natural Philosophy: An Essay with Selections from His Writings,* ed. Marie Boas Hall. Bloomington: Indiana University Press.

Bram, V. A. 1978. "Sentence Construction in Scientific and Engineering Texts." *IEEE Transactions on Professional Communication* 21: 162.

Brannigan, Augustine. 1981. *The Social Basis of Scientific Discoveries.* Cambridge: Cambridge University Press.

Broad, William, and Nicholas Wade. 1982. *Betrayers of the Truth: Fraud and Deceit in the Halls of Science.* New York: Simon and Schuster.

Bronowski, Jacob. 1968. "Honest Jim and the Tinker Toy Model," review of *The Double Helix* by James Watson. *Nation* 206: 381–382.

Brummett, Barry. 1976. "Some Implications of 'Process' and 'Intersubjectivity': Post-Modern Rhetoric." *Philosophy and Rhetoric* 9: 21–51.

Burke, Kenneth. 1962 [1945, 1950]. *A Grammar of Motives and a Rhetoric of Motives.* New York: World.

———. 1969 [1959]. *A Rhetoric of Motives.* Berkeley: University of California Press.

Bytwerk, Randall L. 1979. "The SST Controversy: A Case Study of the Rhetoric of Technology." *Central States Speech Journal* 30: 187–198.

Campbell, John Angus. 1975. "The Polemical Mr. Darwin." *Quarterly Journal of Speech* 61: 375–390.

Carlisle, E. Fred. c. 1983. "Metaphoric Reference in Literature and Science: The Examples of Watson, Crick, and Roethke." 27 pp, unpublished.

Carnap, R. 1963. "Intellectual Autobiography." In *The Philosophy of Rudolph Carnap,* ed. P. A. Schilpp. LaSalle, Ill.: Open Court, pp. 3–84.

Carr, David. 1974. *Phenomenology and the Problem of History.* Evanston, Ill: Northwestern University Press.

Cartwright, Nancy. 1983. *How the Laws of Physics Lie.* Oxford: Clarendon Press.

Cassell's Latin Dictionary. 1955. Revised, J. R. V. Marchant, and J. F. Charles. New York: Funk and Wagnalls.

Chalmers, Alan. 1986. "The Galileo That Feyerabend Missed: An Improved Case against Method." In *The Politics and Rhetoric of Scientific Method: Historical Studies,* ed. John A. Schuster and Richard R. Yeo. Dordrecht: D. Reidel, pp. 1–31.

Chargaff, Erwin. 1974. "Building the Tower of Babble," review of *The Double Helix* by James Watson. *Nature* 248: 776–777.

———. 1976. "On the Dangers of Genetic Meddling." Letter, *Science* 192: 940.

Churchland, Paul M., and Clifford A. Hooker, eds. 1985. *Images of Science: Essays on Realism and Empiricism, with a Reply from Bas C. Van Fraassen.* Chicago: University of Chicago Press.

Clark, Ian D. 1974. "Expert Advice in the Controversy about Supersonic Transport in the United States." *Minerva* 12: 416–432.

Clifford, James, and George E. Marcus, eds. 1986. *Writing Culture: Poetics and Politics of Ethnography.* Berkeley: University of California Press.

Cohen, I. B. 1985. *Revolution in Science.* Cambridge, Mass: Harvard University Press.

———. 1966. *Franklin and Newton: An Inquiry into Speculative Newtonian Experimental Science and Franklin's Work in Electricity as an Example Thereof.* Cambridge, Mass: Harvard University.

Cohen, Morris R., and Ernest Nagel. 1934. *An Introduction to Logic and the Scientific Method.* New York: Harcourt, Brace.

Cole, F. J. 1949. *A History of Comparative Anatomy from Aristotle to the Eighteenth Century.* Reprint, New York: Dover, 1975.

Cole, Jonathan R., and Stephen Cole. 1981. *Peer Review in the National Science Foundation: Phase Two of a Study.* Washington, D.C.: National Academy Press.

———. 1973. *Social Stratification in Science.* Chicago: University of Chicago Press.

Cole, Stephen, Leonard Rubin, and Jonathan R. Cole. 1978. *Peer Review in the National Science Foundation: Phase One of a Study.* Washington, D.C.: National Academy Press.

Collins, H. M. 1975. "The Seven Sexes: A Study in the Sociology of a Phenomenon, or the Replication of Experiments in Physics." *Sociology* 9: 204–224.

Colloquia Copernicana, I. Studia Copernicana, V. 1972. Études sur l'audience de la théorie héliocentrique. Wroclaw [Breslau]: Polska Akademia Nauk.

Cope, Edward Meredith, and John Edward Sandys. 1877. *The Rhetoric of Aristotle with a Commentary,* vol 1. Cambridge: Cambridge University Press.

Copernicus, N. 1978. *On the Revolutions,* ed. J. Dobrzycki, trans. E. Rosen. London: Macmillan.

———. 1959. *Three Copernican Treatises,* 3rd ed., trans. and ed. E. Rosen. New York: Dover.

Crick, F. H. C. 1974. "The Double Helix: A Personal View." *Nature* 248: 766–769.

———. 1954. "The Structure of the Hereditary Material." *Scientific American* 191: 54–61.

Crick, F. H. C., Leslie Barnett, S. Brenner, and R. J. Watts-Tobin. 1961. "General Nature of the Genetic Code for Proteins." *Nature* 192: 1227–1232.

Crombie, A. C. 1959. *Medieval and Early Modern Science,* 2 vols., rev. and enl. ed. New York: Doubleday.

————. 1961. *Robert Grosseteste and the Origins of Experimental Science: 1100–1700*. Oxford: Oxford University Press.

Darwin, Charles. 1958. *The Autobiography of Charles Darwin: 1809–1882,* ed. Nora Barlow. London: Collins.

————. 1987a. *Charles Darwin's Notebooks: 1836–1844,* ed. Paul H. Barrett, Peter J. Gautrey, Sandra Herbert, David Kohn, and Sydney Smith. Ithaca, N.Y.: Cornell University Press.

————. 1977. *The Collected Papers of Charles Darwin,* ed. Paul H. Barrett. Chicago: University of Chicago Press.

————. 1985–. *The Correspondence of Charles Darwin,* ed. Frederick Burkhardt and Sydney Smith. Cambridge: Cambridge University Press.

————. 1963. *Darwin's Ornithological Notes,* ed. Nora Barlow. *Bulletin of the British Museum of Natural History, Historical Series* 2: 201–278.

————. 1987b. *Journal of Researches*. In *The Works of Charles Darwin,* vols. 2 and 3, ed. Paul H. Barrett and R. B. Freeman. New York: New York University Press.

————. 1959 [1887]. *The Life and Letters,* ed. F. Darwin, 2 vols. New York: Basic Books.

————. 1972 [1903]. *More Letters,* ed. F. Darwin, 2 vols. New York: Johnson Reprint.

————. 1964 [1859]. *On the Origin of Species,* 1st ed. Cambridge, Mass.: Harvard University Press.

————. 1962 [1872]. *On the Origin of Species,* 6th ed. New York: Collier.

————. 1980. *The Red Notebook of Charles Darwin,* ed. Sandra Herbert. Ithaca, N.Y.: Cornell University Press.

————. 1962 [1860]. *The Voyage of the Beagle,* ed. Leonard Engel. New York: Doubleday.

David, N. F. 1976. "Crichton's Criticisms." Letter, *Journal of the American Medical Association* 235: 1107.

Davidson, Donald. 1984. *Inquiries into Truth and Interpretation*. Oxford: Clarendon Press.

Davis, Philip J., and Reuben Hersh. 1986. "Mathematics and Rhetoric." In *Descartes' Dream: The World According to Mathematics*. San Diego, Calif.: Harcourt Brace Jovanovich, pp. 57–73.

Day, Robert A. 1979. *How to Write and Publish a Scientific Report*. Philadelphia: ISI Press.

Delgado, Richard, and David R. Millen. 1978. "God, Galileo, and Government: Toward Constitutional Protection for Scientific Inquiry." *Washington Law Journal* 53: 349–404.

De Morgan, A. 1954 [1915]. *A Budget of Paradoxes,* 2nd ed., ed. D. E. Smith. New York: Dover.

Descartes, René. 1954. *Descartes: Philosophical Writings, A Selection,* ed. Elizabeth Anscombe and Peter Thomas Geach. Edinburgh: Thomas Nelson.

————. 1965. *Discourse on Method, Optics, Geometry, and Meteorology,* trans. Paul J. Olscamp. Indianapolis: Bobbs-Merrill.

————. 1979. *Le Monde, ou Traité de la Lumière,* trans. and ed. Michael Sean Mahoney. New York: Abaris.

————. 1898. *Oeuvres de Descartes,* ed. Charles Adam and Paul Tannery. Vol. 2, *Correspondance.* Paris: Léopold Cerf.

————. 1970. *Philosophical Letters,* trans. and ed. Anthony Kenny. Reprint, Minneapolis: University of Minnesota Press, 1981.

————. 1931. *The Philosophical Works,* trans. Elizabeth S. Haldane and G. R. T. Ross, vol. 1. Reprint, Cambridge: Cambridge University Press, 1983.

————. 1983/84. *Principles of Philosophy,* trans. Valentine Rodger Miller and Reese P. Miller. Dordrecht: D. Reidel.

Dickinson, Emily. 1963. *The Poems of Emily Dickinson,* 3 vols., ed. Thomas H. Johnson. Cambridge, Mass.: Harvard University Press.

Douglas, Jack D. 1971. "The Rhetoric of Science and the Origins of Statistical Thought: The Case of Durkheim's *Suicide.*" In *The Phenomenon of Sociology: A Reader in the Sociology of Sociology,* ed. Edward A. Tiryakian. New York: Appleton-Century-Crofts.

Ehrenberg, A. S. C. 1977. "Rudiments of Numeracy." *Journal of the Royal Statistical Society* A, 140: 277–297.

Einstein, Albert. 1959 [1949]. "Autobiographical Notes." In *Albert Einstein: Philosopher-Scientist,* vol. 1, ed. Paul Arthur Schilpp. New York: Harper and Row, pp. 1–95.

————. 1954. *Ideas and Opinions,* trans. Sonja Bargmann. New York: Bonanza Books.

————. 1952 [1923]. *The Principle of Relativity: A Collection of Original Papers on the Special and General Theory of Relativity,* trans. W. Perrett and G. B. Jeffery. New York: Dover.

————. 1961. *Relativity: The Special and General Theory,* trans. Robert W. Lawson. New York: Crown.

————. 1959 [1949]. "Reply to Criticisms." In *Albert Einstein: Philosopher-Scientist,* vol. 2, ed. Paul Arthur Schilpp. New York: Harper and Row, pp. 665–688.

Eldredge, Niles. 1982. Introduction, in Ernst Mayr, *Systematics and the Origin of Species.* New York: Columbia, pp. xv–xxxvii.

Ellis, John M. 1985 [1983]. *One Fairy Story Too Many: The Brothers Grimm and Their Tales.* Chicago: University of Chicago Press.

Erickson, Robert. 1957. *The Structure of Music: A Listener's Guide.* New York: Noonday Press.

Farrell, Thomas B. 1976. "Knowledge, Consensus, and Rhetorical Theory." *Quarterly Journal of Speech* 62: 1–14.

————. 1978. "Social Knowledge II." *Quarterly Journal of Speech* 64: 329–334.

Farrell, Thomas B., and G. Thomas Goodnight. 1981. "Accidental Rhetoric: The Root Metaphors of Three Mile Island." *Communication Monographs* 48: 271–300.

Feyerabend, Paul. 1975. *Against Method.* Reprint, London: Verso, 1978.

————. 1970. "Classical Empiricism." In *The Methodological Heritage of Newton,* ed. Robert E. Butts and John W. Davis. Toronto: University of Toronto Press, pp. 150–170.

Figgis, J. N. 1960 [1916]. *Political Thought from Gerson to Grotius: 1414–1625. Seven Studies*. New York: Harper.

Fine, Arthur. 1984. "The Natural Ontological Attitude." In *Scientific Realism*, ed. Jarrett Leplin. Berkeley: University of California Press, pp. 83–107.

———. 1986. *The Shaky Game: Einstein, Realism, and the Quantum Theory*. Chicago: University of Chicago Press.

Fish, Stanley E. 1974 [1972]. *Self-Consuming Artifacts: The Experience of Seventeenth-Century Literature*. Berkeley: University of California Press.

Fitzpatrick, J. W. 1980. "A New Race of *Atlapetes leucopterus*, with Comments on Widespread Albinism in *A. l. dresseri* (Taczanowski)." *Auk* 97: 883–887.

Fitzpatrick, J. W., and J. P. O'Neill. 1979. "A New Tody-Tyrant from Northern Peru." *Auk* 96: 443–447.

Fitzpatrick, J. W., D. E. Willard, and J. W. Terborgh. 1979. "A New Species of Hummingbird from Peru." *The Wilson Bulletin* 91: 177–186.

Fleck, Ludwik. 1979 [1935]. *Genesis and Development of a Scientific Fact*, ed. Thaddeus J. Trenn and Robert K. Merton, trans. Fred Bradley and Thaddeus J. Trenn. Chicago: University of Chicago Press.

Fodor, Jerry A. 1975. *The Language of Thought*. Cambridge, Mass.: Harvard University Press.

Frank, David A. 1981. "'Shalom Achsav': Rituals of the Israeli Peace Movement." *Communication Monographs* 48: 165–182.

Franklin, Allan. 1986. *The Neglect of Experiment*. Cambridge: Cambridge University Press.

Freud, Sigmund. 1949 [1940]. *An Outline of Psychoanalysis*, trans. James Strachey. New York: W. W. Norton.

F.R.S. 1968. "Notes of a Not-Watson," review of *The Double Helix* by James Watson. *Encounter* 31 (July 1968): 60–66.

Gadamer, Hans-Georg. 1975 [1965]. *Truth and Method*, trans. Garret Barden and John Cumming. New York: Crossroad.

Galilei, Galileo. 1957. *Discoveries and Opinions of Galileo*, ed. Stillman Drake. New York: Doubleday.

Galison, Peter. 1983. "Rereading the Past from the End of Physics." In *Functions and Uses of Disciplinary Histories*, ed. Loren Graham, Wolf Lepenies, and Peter Weingart. Dordrecht: D. Reidel, pp. 35–51.

Gallie, W. B. 1968 [1964]. *Philosophy and the Historical Understanding*, 2nd ed. New York: Schocken Books.

Gaukroger, Stephen. 1980. "Descartes' Project for a Mathematical Physics." In *Descartes: Philosophy, Mathematics and Physics*, ed. Stephen Gaukroger. Sussex: Harvester Press.

Geertz, Clifford. 1973. *The Interpretation of Cultures*. New York: Basic Books.

———. 1983. *Local Knowledge: Further Essays in Interpretive Anthropology*. New York: Basic Books.

Gell-Mann, M., and F. E. Low. 1954. "Quantum Electrodynamics at Small Distances." *Physical Review* 2nd series, 95 (September 1, 1954): 1300–1317.

General Notes on the Preparation of Scientific Papers. 1950. 1st ed. London: Royal Society.

General Notes on the Preparation of Scientific Papers. 1965. 2nd ed. London: Royal Society.

Ghiselin, Michael T. 1984. *The Triumph of the Darwinian Method.* Chicago: University of Chicago Press.

Gieryn, Thomas F. 1983. "Boundary-Work and the Demarcation of Science from Non-Science: Strains and Interests in Professional Ideologies of Scientists." *American Sociological Review* 48: 781–795.

Gilbert, G. Nigel. 1976. "The Transformation of Research Findings into Scientific Knowledge." *Social Studies of Science* 6: 281–306.

Gilbert, G. Nigel, and Michael Mulkay. 1984. *Opening Pandora's Box: A Sociological Analysis of Scientists' Discourse.* Cambridge: Cambridge University Press.

Gillispie, Charles Coulston. 1973 [1960]. *The Edge of Objectivity: An Essay in the History of Scientific Ideas.* Princeton, N.J.: Princeton University Press.

Goodman, Nelson. 1972. *Problems and Projects.* Indianapolis: Bobbs-Merrill.

———. 1978. *Ways of Worldmaking.* Indianapolis: Hackett.

Goudsmit, S. A. 1967. Letter, *Physics Today* 20 (January 1967): 12.

Graham, Loren R. 1980. "Reasons for Studying Soviet Science: The Example of Genetic Engineering." In *The Social Context of Soviet Science*, ed. Linda L. Lubrano and Susan Gross Solomon. Boulder, Colo.: Westview Press, pp. 205–240.

Graham, Loren, Wolf Lepenies, and Peter Weingart, eds. 1983. *Functions and Uses of Disciplinary Histories.* Dordrecht: D. Reidel.

Grant, E. 1962. "Late Medieval Thought, Copernicus, and the Scientific Revolution." *Journal of the History of Ideas* 23: 197–220.

Green, Arnold L. 1961. "The Ideology of Anti-Fluoridation Leaders." *Journal of Social Issues* 17: 13–25.

Greenfield, D. W., and G. S. Glodek. 1977. "*Trachelyichthys exilis*, A New Species of Catfish (Pisces: Auchenipteridae) from Peru." *Fieldiana: Zoology* 72: 47–58.

Grimm, Jacob, and Wilhelm Grimm. 1980 [1957]. *Die Märchen.* Munich: Wilhelm Goldmann Verlag.

Grobstein, Clifford. 1979. *A Double Image of the Double Helix: The Recombinant DNA Debate.* San Francisco: W. H. Freeman.

Gross, Alan G. 1988. "Adaptation in Evolutionary Epistemology: Clarifying Hull's Model." *Biology and Philosophy* 3: 185–186.

———. 1983a. "Analogy and Intersubjectivity in Political Oratory, Scholarly Argument, and Scientific Reports." *Quarterly Journal of Speech* 69: 37–46.

———. 1985. "The Form of the Experimental Paper: A Realization of the Myth of Induction." *Journal of Technical Writing and Communication* 15: 15–26.

————. 1983b. "A Primer on Tables and Figures." *Journal of Technical Writing and Communication* 13: 33–55.

————. 1984. "Style and Arrangement in Scientific Prose: The Rules behind the Rules," *Journal of Technical Writing and Communication* 14: 241–253.

————. 1987. "A Tale Twice Told: The Rhetoric of Discovery in the Case of DNA." In *Argument and Critical Practices: Proceedings of the Fifth SCA/AFA Conference on Argumentation,* 1987 ed. Joseph W. Wenzel. Annandale, Va.: Speech Communication Association, pp. 491–498.

Gruber, Howard E. 1981. *Darwin on Man: A Psychological Study of Scientific Creativity,* 2nd ed. Chicago: University of Chicago Press.

————. 1985. "Going the Limit: Toward the Construction of Darwin's Theory (1832–1839)." In *The Darwinian Heritage,* ed. David Kohn. Princeton, N.J.: Princeton University Press, pp. 9–34.

Guerlac, Henry. 1981. *Newton on the Continent.* Ithaca, N.Y.: Cornell University Press.

Gusfield, Joseph. 1976. "The Literary Rhetoric of Science: Comedy and Pathos in Drinking Driver Research." *American Sociological Review* 41: 16–34.

Habermas, Jürgen. 1979 [1976]. *Communication and the Evolution of Society,* trans. Thomas McCarthy. Boston: Beacon Press.

————. 1971 [1968]. *Knowledge and Human Interests,* trans. J. J. Shapiro. Boston: Beacon Press.

————. 1982. "A Reply to My Critics." In *Habermas: The Critical Debates,* ed. John B. Thompson and David Held. Cambridge, Mass.: MIT Press, pp. 219–283.

————. 1984. *The Theory of Communicative Action,* vol. 1, *Reason and the Rationalization of Society,* trans. Thomas McCarthy. Boston: Beacon Press.

————. 1987. *The Theory of Communicative Action,* vol. 2, *Lifeworld and System: A Critique of Functionalist Reason,* trans. Thomas McCarthy. Boston: Beacon Press.

————. 1973. "Wahrheitstheorien." In *Wirklichkeit und Reflexion: Festschrift für Walter Schulz.* Pfüllingen: Gunther Neske, pp. 211–265.

Hacking, Ian. 1987. "Extragalactic Reality: The Case of Gravitational Lensing." Notes for the Newton and Scientific Realism Workshop (unpublished), Van Leer Institute, Jerusalem, April 27–30, 1987, 47 pp.

————. 1986. "The Making and Molding of Child Abuse: An Exercise in Describing a Kind of Human Behavior." Harris Lecture (unpublished), Northwestern University, May 7, 1986, 60 pp.

————. 1983. *Representing and Intervening: Introductory Topics in the Philosophy of Natural Science.* Cambridge: Cambridge University Press.

Hall, A. Rupert. 1980. *Philosophers at War: The Quarrel between Newton and Leibniz.* Cambridge: Cambridge University Press.

Halloran, S. Michael. 1980. "Toward a Rhetoric of Scientific Revolution." In *Proceedings: 31st Conference on College Composition and Communication: Technical Communication Sessions,* ed. John A. Muller. Urbana, Ill.: ATTW, pp. 229–236.

Handbook for Authors of Reports in American Chemical Society Publications. 1978. Washington, D.C.: American Chemical Society.

Harré, Rom. 1984. *Personal Being: A Theory for Individual Psychology.* Cambridge, Mass.: Harvard University Press.

Hart, H. L. A. 1980. "Definition and Theory of Jurisprudence." In *Philosophy of Law*, 2nd ed., ed. Joel Feinberg and Hyman Gross. Belmont, Calif.: Wadsworth, pp. 252–258.

Hawthorne, Nathaniel. 1964. "Dr. Rappaccini's Daughter." In *Selected Tales and Sketches*, 3rd ed., ed. Hyatt H. Waggoner. New York: Holt.

Hayles, N. Katherine. 1984. *The Cosmic Web: Scientific Field Models and Literary Strategies in the Twentieth Century.* Ithaca, N.Y.: Cornell University Press.

———. 1987. "The Politics of Chaos: Local Knowledge versus Global Theory." Conference on Argument in Science: New Sociologies of Science/Rhetoric of Inquiry, Iowa City, October 9–11, 1987.

Held, David. 1980. *Introduction to Critical Theory.* Berkeley: University of California Press.

Herbert, Sandra. 1974; 1977. "The Place of Man in the Development of Darwin's Theory of Transmutation," parts 1 and 2. *Journal of the History of Biology* 7 (1974): 217–258; 10 (1977): 155–227.

Hesse, Mary B. 1966. *Models and Analogies in Science.* Notre Dame, Ind.: University of Notre Dame Press.

Holdsworth, Sir William. 1966 [1925]. *A History of English Law*, 16 vols. London: Methuen.

Hollis, M., and S. Lukes, eds. 1982. *Rationality and Relativism.* Cambridge, Mass.: MIT Press.

Hooke, Robert. 1665. *Micrographia, or Some Physiological Descriptions of Minute Bodies Made by Magnifying Glasses with Observations and Inquiries Thereupon.* Reprint, New York: Dover, 1938.

Hugonnard-Roche, Henri, and Jean-Pierre Verdet, eds. 1982. *Narratio Prima.* In *Studia Copernica*, 20.

Hull, D. L. 1983a. "Darwin and the Nature of Science." In *Evolution from Molecules to Men*, ed. D. S. Bendall. Cambridge: Cambridge University Press, pp. 63–80.

———. 1983b. "Karl Popper and Plato's Metaphor." In *Advances in Cladistics*, vol. 2, ed. N. Platnick and V. Funk. New York: Columbia University Press, pp. 177–189.

———. 1984 [1978]. "A Matter of Individuality." In *Conceptual Issues in Evolutionary Biology: An Anthology*, ed. E. Sober. Cambridge, Mass: MIT Press, pp. 623–645.

———. 1988. "A Mechanism and Its Metaphysics: An Evolutionary Account of the Social and Conceptual Development of Science." *Biology and Philosophy* 3: 123–155.

———. 1981a. "The Principles of Biological Classification: The Use and Abuse of Philosophy." *PSA 1976* 2: 130–153.

———. 1981b. "Reduction and Genetics." *Journal of Medicine and Philosophy* 6: 125–143.

Husserl, Edmund. 1970. *The Crisis of European Sciences and Transcendental Phenomenology*, trans. David Carr. Evanston, Ill.: Northwestern University Press.

Huygens, Christian. 1888–1950. *Oeuvres Complètes de Christian Huygens publiées par la Société Hollandaise des Sciences*, 22 vols. The Hague: Martinus Nijhoff.

"Inadmissible Evidence." 1981. *Scientific American* 245 (1981): 78.

"Instructions for Contributors." 1983. *Science* 222 (1983): xi–xii.

Ivins, William M., Jr. 1938. *On the Rationalization of Sight, with an Examination of Three Renaissance Texts on Perspective*. Papers, no. 8. New York: The Metropolitan Museum of Art.

Jardine, Lisa. 1974. *Francis Bacon: Discovery and the Art of Discourse*. Cambridge: Cambridge University Press.

Jenks, Edward. 1949 [1912]. *A Short History of English Law from the Earliest Times to the End of the Year 1938*. London: Methuen.

Johnson-Laird, P. N. 1968. "The Choice of the Passive Voice in a Communicative Task." *British Journal of Psychology* 59: 7–15.

——. 1968. "The Interpretation of the Passive Voice." *Quarterly Journal of Experimental Psychology* 20: 69–73.

——. 1977. "The Passive Paradox: A Reply To Costermans and Hupet." *British Journal of Psychology* 68: 113–116.

Johnstone, Henry W., Jr. 1963. "Can Philosophical Argument Be Valid?" *Bucknell Review* 11: 89–98.

——. 1978. *Validity and Rhetoric in Philosophical Argument: An Outlook in Transition*. University Park, Pa.: The Dialogue Press of Man and World.

Jones, Richard Foster. 1961 [1936]. *Ancients and Moderns: A Study of the Rise of the Scientific Movement in Seventeenth-Century England*. New York: Dover.

Karon, L. A. 1976. "Presence in *The New Rhetoric*." *Philosophy and Rhetoric* 9: 96–111.

Keller, Evelyn Fox. 1985. *Reflections on Gender and Science*. New Haven: Yale University Press.

Kelso, James A. 1980. "Science and the Rhetoric of Reality." *Central States Speech Journal* 31: 17–29.

Kepler, J. 1981. *Mysterium Cosmographicum. The Secret of the Universe*, trans. A. M. Duncan. New York: Abaris.

Kethley, J. 1983. "The Deutonymph of *Epiphis rarior* Berlese, 1916 (Epiphidinae n. subfam., Rhodacaridae, Rhodacaroidea)." *Canadian Journal of Zoology* 61: 2598–2611.

King, L. S. 1978. "Better Writing Anyone?" Letter, *JAMA* 239: 752.

Kinneavy, James L. 1971. *A Theory of Discourse*. New York: W. W. Norton.

Klatzky, Roberta. 1980. *Human Memory: Structure and Processes*, 2nd ed. San Francisco: W. H. Freeman.

Knorr-Cetina, Karin D. 1981. *The Manufacture of Knowledge: An Essay on the Constructivist and Contextual Nature of Science*. Oxford: Pergamon Press.

Koestler, A. 1968 [1959]. *The Sleepwalkers*. New York: Macmillan.

Köhler, Wolfgang. 1947. *Gestalt Psychology*. New York: New American Library.

Kohn, David. 1980. "Theories to Work By: Rejected Theories, Reproduction, and Darwin's Path to Natural Selection," *Studies in History of Biology* 4: 67–170.

Kolata, Gina Bari. 1981. "Reevaluation of Cancer Data Eagerly Awaited." *Science* 214: 316–318.

Knappen, M. M. 1964. *Constitutional and Legal History of England*. Haden, Conn.: Archon Books.

Koyré, Alexandre. 1973 [1961]. *The Astronomical Revolution: Copernicus—Kepler—Borelli*, trans. R. E. W. Maddison. Paris: Hermann.

———. 1968. *Newtonian Studies*. Chicago: University of Chicago Press.

Kronick, David A. 1976. *A History of Scientific and Technical Periodicals: The Origins and Development of the Scientific and Technical Press: 1665–1790*, 2nd ed. Metuchin, N.J.: The Scarecrow Press.

Kuhn, Thomas S. 1981 [1957]. *The Copernican Revolution: Planetary Astronomy in the Development of Western Thought*. Cambridge, Mass.: Harvard University Press.

———. 1977. *The Essential Tension: Selected Studies in Scientific Tradition and Change*. Chicago: University of Chicago Press.

———. 1970. "Reflections on My Critics." In *Criticism and the Growth of Knowledge*, ed. Imre Lakatos and Alan Musgrave. Cambridge: Cambridge University Press, pp. 231–278.

———. 1962. *The Structure of Scientific Revolutions*. Chicago: University of Chicago Press.

Lakatos, Imre. 1983 [1972–73]. "Why Did Copernicus' Research Programme Supersede Ptolemy's?" In *The Methodology of Scientific Research Programmes: Philosophical Papers*, vol. 1, ed. J. Worrall and G. Currie. Cambridge: Cambridge University Press, pp. 168–192.

Latour, Bruno. 1987. *Science in Action*. Cambridge, Mass.: Harvard University Press.

———. 1986. "Visualization and Cognition: Thinking with Eyes and Hands." In *Knowledge and Society: Studies in the Sociology of Culture Past and Present*, ed. Henrika Kuklick and Elizabeth Long. Greenwich, Conn.: Jai Press, pp. 1–40.

Latour, Bruno, and Jocelyn de Noblet, eds. 1985. "Les 'Vues' de L'Espirit." *Culture Technique* 14.

Latour, Bruno, and Steve Woolgar. 1979. *Laboratory Life: The Social Construction of Scientific Facts*. Sage Library of Social Research, vol. 80. Beverly Hills, Calif.: Sage Publications.

Laudan, Larry. 1984. "A Confutation of Convergent Realism." In *Scientific Realism*, ed. Jarrett Leplin. Berkeley: University of California Press.

Laudan, Rachel. 1983. "Redefinitions of a Discipline: Histories of Geology and Geological History." In *Functions and Uses of Disciplinary Histories*, ed. Loren Graham, Wolf Lepenies, and Peter Weingart. Dordrecht: D. Reidel, pp. 79–104.

Laymon, Ronald. 1984. "The Path from Data to Theory." In *Scientific Realism*, ed. Jarrett Leplin. Berkeley: University of California Press, pp. 108–123.

Leach, Edmund. 1980 [1974]. *Claude Lévi-Strauss*, rev. ed. Harmondsworth, Middlesex: Penguin Books.

———. 1982 [1976]. *Culture and Communication: The Logic by Which Symbols Are Connected*. Cambridge: Cambridge University Press.

Lear, John. 1978. *Recombinant DNA: The Untold Story*. New York: Crown.

Leeuwenhoek, Antony van. 1960 [1932]. *Antony van Leeuwenhoek and his "Little Animals,"* ed. Clifford Dobell. New York: Dover.

Leibniz, Gottfried Wilhelm. 1976 [1969]. *Philosophical Reports and Letters*, 2nd ed., ed. Leroy E. Loemker. Dordrecht: D. Reidel.

Leuchtenburg, William E. 1973. *Franklin D. Roosevelt and the New Deal: 1932–40*. New York: Harper and Row.

Lévi-Strauss, Claude. 1963. *Structural Anthropology*, trans. Claire Jacobson and Brooke Grundfest Schoepf. New York: Basic Books.

———. 1976. *Structural Anthropology, 2*, trans. Monique Layton. Chicago: University of Chicago Press.

Lewontin, Richard C. "'Honest Jim's'" Big Thriller about DNA (1968)," review of *The Double Helix* by James Watson. *Chicago Sunday Sun-Times*, February 25, 1968, pp. 1–2.

———. 1984. "The Structure of Evolutionary Genetics." In *Conceptual Issues in Evolutionary Biology: An Anthology*, ed. E. Sober. Cambridge: MIT Press, pp. 3–13.

Limon, John. 1986. "*The Double Helix* as Literature." *Raritan* 5: 26–47.

Lin, N. 1972. "A Comparison between the Scientific Communication Model and the Mass Communication Model: Implications for the Transfer and Utilization of Scientific Knowledge." *IEEE Transactions on Professional Communication* 15: 34–38.

Lindberg, David C. 1968. "The Cause of Refraction in Medieval Optics." *British Journal for the History of Science* 4: 23–38.

———. 1976. *Theories of Vision from Al-Kindi to Kepler*. Chicago: University of Chicago Press.

Lloyd, G. E. R. 1971. *Polarity and Analogy: Two Types of Argumentation in Early Greek Thought*. Cambridge: Cambridge University Press.

Locke, John. 1979 [1975]. *An Essay Concerning Human Understanding*, ed. Peter H. Nidditch. Oxford: The Clarendon Press.

Luria, S. F. 1986. Review of *The Transforming Principle* by Maclyn McCarty. *Scientific American* 254 (April 1986): 24–31.

Lwoff, André. 1968. Review of *The Double Helix* by James Watson. *Scientific American* 219 (July 1968): 132–138.

Lyell, Charles. 1969 [1830–33]. *Principles of Geology*, 1st ed., 3 vols., intro. by Martin J. S. Rudwick. New York: Johnson Reprint Corporation.

Lynch, Michael. 1985a. *Art and Artifact in Laboratory Science: A Study of Shop Work and Shop Talk in a Research Laboratory*. London: Routledge and Kegan Paul.

———. 1985b. "Discipline and the Material Form of Images: An Analysis of Scientific Visibility." *Social Studies of Science* 15: 37–66.

Lyons, John. 1978 [1977]. *Semantics*, 2 vols. Cambridge: Cambridge University Press.

McCarthy, Thomas. 1982a. *The Critical Theory of Jürgen Habermas*. Cambridge, Mass.: MIT Press.

————. 1982b. "Rationality and Relativism: Habermas' 'Overcoming' of Hermeneutics." In *Habermas: The Critical Debates*, ed. J. B. Thompson and D. Held. Cambridge, Mass.: MIT Press, pp. 57–78.

————. 1973. "A Theory of Communicative Competence." *Philosophy of the Social Sciences* 3: 135–156.

McKeon, Richard. 1949. "Aristotle and the Origins of Science in the West." In *Science and Civilization*, ed. Robert C. Stauffer. Madison: University of Wisconsin Press, pp. 3–29.

Maestlin, Michael. 1938. Appendix to Kepler's *Mysterium Cosmographicum*, ed. Max Caspar. In J. Kepler, *Gesammelte Werke*, vol 1. Munich: C. H. Beck'sche.

Mahon, B. H. 1977. "Statistics and Decisions: The Importance of Communication and the Power of Graphical Presentation." *Journal of the Royal Statistical Society* 140: 298–323.

Manier, Edward. 1978. *The Young Darwin and His Cultural Circle: A Study of Influences Which Helped Shape the Language and Logic of the Theory of Natural Selection*. Dordrecht: D. Reidel.

Manuel, Frank E. 1979 [1968]. *A Portrait of Isaac Newton*. Washington, D.C.: New Republic Books.

Markus, Gyorgy. 1987. "Why Is There No Hermeneutics of Natural Sciences? Some Preliminary Theses." *Science in Context* 1: 5–51.

Marlowe, Christopher. 1910. *Dr. Faustus*. In the *Works*, ed. F. Tucker Brooke. Oxford: The Clarendon Press.

Mayr, Ernst. 1982 [1942]. *Systematics and the Origin of Species*. New York: Columbia University Press.

Mazuzan, George T. 1982. "Atomic Power Safety: The Case of the Power Reactor Development Company Fast Breeder, 1955–56." *Technology and Culture*, 23: 341–371.

Mazur, Allan. 1973. "Disputes between Experts." *Minerva* 11: 243–262.

Medawar, Peter. 1964. "Is the Scientific Report Fraudulent? Yes; It Misrepresents Scientific Thought." *Saturday Review* 47 (August 1, 1964): 42–43.

————. 1968. "Lucky Jim," review of *The Double Helix* by James Watson. *New York Review of Books* (March 28, 1968): 3–5.

————. 1984. *Pluto's Republic*. Oxford: Oxford University Press.

Merton, Robert K. 1973. *The Sociology of Science: Theoretical and Empirical Investigations*, ed. Norman W. Storer. Chicago: University of Chicago Press.

————. 1987. "Three Fragments from a Sociologist's Notebooks: Establishing the Phenomenon, Specified Ignorance, and Strategic Research Materials." *Annual Review of Sociology* 13: 1–28.

Moesgaard, K. P. 1972. "Copernican Influence on Tycho Brahe." In *Colloquia Copernicana 1. Studia Copernicana V*, Études sur l'audience de la Théorie Héliocentrique. Wroclaw [Breslau]: Polska Akademia Nauk, pp. 31–55.

Montalbo, Thomas. 1978. "Winston Churchill: A Study in Oratory." *IEEE Transactions on Professional Communications* 21: 5–8.

Morgan, Joan, and W. J. Whelan, eds. 1979. *Recombinant DNA and Genetic Experimentation*. Proceedings of the Conference on Recombinant DNA, jointly organized by the Committee on Genetic Experimentation (COGENE) and the Royal Society of London, held at Wye College, Kent, April 1–4, 1979. Oxford: Pergamon Press.

Mulkay, Michael, and Nigel Gilbert. 1981. "Putting Philosophy to Work: Karl Popper's Influence on Scientific Practice." *Philosophy of the Social Sciences* 11: 389–407.

Myers, Greg. 1985. "Texts as Knowledge Claims: The Social Construction of Two Biology Articles." *Social Studies of Science* 15: 593–630.

Nahm, Milton C., ed. 1964. *Selections from Early Greek Philosophy*. New York: Appleton-Century-Crofts.

National Academy of Sciences. 1977. *Research with Recombinant DNA: An Academy Forum*. March 7–9, 1977. Washington, D.C.: National Academy of Sciences.

Neugebauer, O. 1970 [1957]. *The Exact Sciences in Antiquity*, 2nd ed. Providence, R.I.: Brown University Press.

Newton, Isaac. 1980 [1715]. "Account of the Book Entituled *Commercium Epistolicum*." *Philosophical Transactions* 29 (1715): 173–224. Reprinted in A. Rupert Hall, *Philosophers at War: The Quarrel between Leibniz and Newton*. Cambridge: Cambridge University Press, pp. 263–314.

———. 1959–1977. *The Correspondence*, ed. H. W. Turnbull, J. F. Scott, A. Rupert Hall, and Laura Tilling, 7 vols. Cambridge: Cambridge University Press.

———. 1978. *Isaac Newton's Papers and Letters on Natural Philosophy*, 2nd ed., ed. I. Bernard Cohen and Robert E. Schofield. Cambridge, Mass.: Harvard University Press.

———. 1969. *The Mathematical Papers*, vol 3, ed. D. T. Whiteside, M. A. Hoskin, and A. Prag. Cambridge: Cambridge University Press.

———. 1974 [1934]. *Mathematical Principles of Natural Philosophy and His System of the World, Translated into English by Andrew Motte in 1729*, ed. Florian Cajori, 2 vols. Berkeley: University of California Press.

———. 1730. *Opticks, or a Treatise of the Reflections, Refractions, Inflections, and Colours of Light*, 4th ed. Reprint, New York: Dover, 1979.

Nirenberg, Marshall W., and J. Heinrich Matthaei. 1961. "The Dependence of Cell-Free Protein Synthesis in *E. Coli* upon Naturally Occurring or Synthetic Polyribonucleotides." *Proceedings of the National Academy of Sciences* 47: 1588–1602.

Oakeshott, Michael. 1962. *Rationalism in Politics and Other Essays*. New York: Basic Books.

Olby, Robert. 1974. *The Path to the Double Helix*. Seattle: University of Washington Press.

Oldenburg, Henry. 1965–1973. *The Correspondence*, ed. A. Rupert Hall and

Marie Boas Hall, 11 vols. Madison: University of Wisconsin Press; London: Mansell, 1965–1977.

"Oncogenes." 1981. *Scientific American* 244: 90–93.

Ortega y Gasset, José. 1960 [1930]. *The Revolt of the Masses.* New York: W. W. Norton.

Osborn, Michael. 1967. "Archetypal Metaphor in Rhetoric: The Light-Dark Family." *Quarterly Journal of Speech* 53: 115–126.

Overington, Michael. 1977. "The Scientific Community as Audience." *Philosophy and Rhetoric* 10: 111–121.

Oxford Dictionary of Proverbs. 1970. 3rd ed., rev. F. P. Wilson, Oxford: The Clarendon Press.

Patterson, B. D. 1982. "Pleistocene Vicariance, Montane Islands, and the Evolutionary Divergence of Some Chipmunks (Genus *Eutamias*)." *Journal of Mammology* 63: 387–398.

Patterson, C. 1982. "Cladistics." In *Evolution Now: A Century after Darwin*, ed. J. M. Smith. San Francisco: W. H. Freeman, pp. 110–120.

Pauling, Linus. 1974. "Molecular Basis of Biological Specificity." *Nature* 248: 769–771.

———. 1952. In *Les Protéins: Rapport et Discussions*, Neuvième Conseil de Chimie. Brussels: Institute International de Chimie Solvay.

Pauling, Linus, Robert B. Corey, and H. R. Branson. 1951. "The Structure of Proteins: Two Hydrogen-Bonded Helical Configurations of the Polypeptide Chain." *National Academy of Science: Proceedings* 37: 205–211.

Pauling, Peter. 1973. "DNA—The Race That Never Was." *New Scientist* (May 31, 1973): 558–560.

Peer Commentary on Peer Review: A Case Study in Scientific Quality Control. 1982. Reprinted from *The Behavioral and Brain Sciences*, ed. Stevan Harnad. Cambridge: Cambridge University Press.

Peirce, Charles. 1955. *Philosophical Writings of Peirce*, ed. Justus Buchler. New York: Dover.

Perelman, Chaim, and L. Obrechts-Tyteca. 1971 [1958]. *The New Rhetoric: A Treatise on Argumentation*, trans. John Wilkinson and Purcell Weaver. Notre Dame, Ind.: University of Notre Dame Press.

Perutz, Max. 1969. Letter, *Science* 164: 1537–1538.

Pickering, Andrew. 1984. *Constructing Quarks: A Sociological History of Particle Physics.* Chicago: University of Chicago Press.

Pinch, Trevor. 1985a. "Theory Testing in Science—The Case of the Solar Neutrinos: Do Crucial Experiments Test Theories or Theorists?" *Philosophy of the Social Sciences* 15: 167–187.

Pinch, Trevor. 1985b. "Towards an Analysis of Scientific Observation: The Externality and Evidential Significance of Observational Reports in Physics." *Social Studies in Science* 15: 3–36.

Polanyi, Michael. 1964. *Science, Faith, and Society.* Chicago: University of Chicago Press.

Pollock, Sir Frederick, and Frederick William Maitland. 1968 [1895]. *The History of English Law before the Time of Edward I*, 2 vols., 2nd ed. Cambridge: Cambridge University Press.

Polya, G. 1954. *Of Mathematics and Plausible Reasoning*, vol. 1, *Induction and Analogy in Mathematics*. Princeton, N.J.: Princeton University Press.

Popper, K. R. 1965. *Conjectures and Refutations: The Growth of Scientific Knowledge*. New York: Harper and Row.

——. 1968 [1934]. *The Logic of Scientific Discovery*. New York: Harper and Row.

——. 1970. "Normal Science and Its Dangers." In *Criticism and the Growth of Knowledge*, ed. Imre Lakatos and Alan Musgrave. Cambridge: Cambridge University Press, pp. 50–58.

Portugal, Franklin H., and Jack S. Cohen. 1977. *A Century of DNA: A History of the Discovery of the Structure and Function of the Genetic Substance*. Cambridge, Mass.: MIT Press.

Propp, V. 1984 [1968]. *Morphology of the Folktale*, 2nd ed., trans. Laurence Scott. Austin: University of Texas Press.

Prowe, L., ed. 1967 [1883–84]. *Nicolaus Coppernicus*, vol. 2. Osnbrück: Otto Zeller.

Putnam, Hilary. 1987. *The Many Faces of Realism*. LaSalle, Ill.: Open Court.

——. 1986 [1981]. *Reason, Truth, and History*. Cambridge: Cambridge University Press.

Quine, Willard Van Orman. 1969. *Ontological Relativity and Other Essays*. New York: Columbia University Press.

——. 1961 [1953]. "On What There Is." In *From a Logical Point of View: Logico-Philosophical Essays*. New York: Harper and Row, pp. 1–19.

——. 1970. *Philosophy of Logic*. Englewood Cliffs, N.J.: Prentice-Hall.

——. 1976 [1966]. *The Ways of Paradox and Other Essays*, rev. and enl. ed. Cambridge, Mass.: Harvard University Press.

——. 1960. *Word & Object*. Cambridge, Mass.: MIT Press.

Quintilian. 1920–1922. *Institutio Oratoria*, trans. H. E. Butler, 4 vols. Cambrige, Mass.: Harvard University Press.

Quirk, Randolph, Sidney Greenbaum, Geoffrey Leech, and Jan Svartvik. 1979. *A Grammar of Contemporary English*. London: Longman.

Racker, Efraim. 1983. "The Warburg Effect: Two Years Later." Letter, *Science* 222: 232.

——. 1981. "Warburg Effect Revisited." Letter, *Science* 213: 1313.

Racker, Efraim, and Mark Spector. 1981. "Warburg Effect Revisited: Merger of Biochemistry and Molecular Biology." *Science*, 213: 303–307.

Recker, Doren A. 1987. "Causal Efficacy: The Structure of Darwin's Argument Strategy in the *Origin of Species*." *Philosophy of Science* 54: 147–176.

"Recombinant DNA Research: A Debate on the Benefits and Risks." 1977. *Chemical and Engineering News* 55 (May 30, 1977): 26–42.

Rephaeli, Ada, Mark Spector, and Efraim Racker. 1981. "Stimulation of Ca^{2+} Uptake and Protein Phosphorylation in Tumor Cells by Fibronectin." *Journal of Biological Chemistry* 256: 6069–6074.

Rheticus. 1959. *Narratio Prima*. In *Three Copernican Treatises*, 2nd ed., ed. and trans. E. Rosen. New York: Dover.

——. 1982. *Narratio Prima*, ed. Henri Hugonnard-Roche and Jean-Pierre Verdet. *Studia Copernica* 20.

Richards, Robert J. 1987. *Darwin and the Emergence of Evolutionary Theories of Mind and Behavior*. Chicago: University of Chicago Press.

Ricqlès, A. de, and J. R. Bolt. 1983. "Jaw Growth and Tooth Replacement in *Captorhinus Aguti* (Reptilia: Captorhinomorpha): A Morphological and Histological Analysis." *Journal of Vertibrate Paleontology* 3: 7–24.

Rogers, Michael. 1977. *Biohazard*. New York: Knopf.

Rose, Steven. 1987. *Molecules and Minds: Essays on Biology and the Social Order*. Milton Keynes, England: Open University Press.

Roosevelt, Franklin D. 1972. *Complete Presidential Press Conferences of Franklin D. Roosevelt, Vols. 1–2: 1933*. New York: DaCapo Press.

———. 1963. "First Inaugural Address." *Famous Speeches in American History*, ed. Glenn A. Capp. Indianapolis: Bobbs-Merrill, pp. 193–198.

Rosen, E., ed. and trans. 1959. *Narratio Prima*. In *Three Copernican Treatises*, 2nd ed. New York: Dover.

Ross, D. 1971 [1949]. *Aristotle*. London: Methuen.

Rowland, R. C. 1982. "The Influence of Purpose on Fields of Argument." *Journal of the American Forensic Association* 18: 228–245.

Russell, Bertrand. 1974. "On Induction." Reprinted in *The Justification of Induction*, ed. Richard Swinburne. London: Oxford University Press, pp. 1–25.

Ryle, Gilbert. 1949. *The Concept of Mind*. New York: Barnes and Noble.

Sayre, Anne. 1975. *Rosalind Franklin and DNA*. New York: Norton.

Schattschneider, E. E. 1975. *The Semisovereign People: A Realist's View of Democracy in America*. Hinsdale, Ill.: Dryden Press.

Schlesinger, Arthur M. 1959. *The Age of Roosevelt: The Coming of the New Deal*. Boston: Houghton-Mifflin.

Schrödinger, Erwin. 1967. *What is Life? The Physical Aspect of the Living Cell*. Cambridge: Cambridge University Press.

Schuster, John A. 1986. "Cartesian Method as Mythic Speech: A Diachronic and Structural Analysis." In *The Politics and Rhetoric of Scientific Method: Historical Studies*, ed. John A. Schuster and Richard R. Yeo. Dordrecht: D. Reidel, pp. 33–95.

Schuster, John A., and Richard R. Yeo, eds. 1986. *The Politics and Rhetoric of Scientific Method: Historical Studies*. Dordrecht: D. Reidel.

Sciama, D. W. 1959. *The Physical Foundations of General Relativity*. New York: Doubleday.

Scruton, Roger. 1982. *Kant*. Oxford: Oxford University Press.

Searle, John R. 1969. *Speech Acts: An Essay in the Philosophy of Language*. Cambridge: Cambridge University Press.

Shapin, S. 1984. "Pump and Circumstance: Robert Boyle's Literary Technology." *Social Studies of Science* 14: 481–520.

Shelley, Mary. 1963. *Frankenstein*. London: Dent.

Shweder, Richard A. 1986. "Divergent Rationalities." In *Metatheory in Social Science: Pluralisms and Subjectivities*, ed. Donald W. Fiske and Richard A. Shweder. Chicago: University of Chicago Press, pp. 163–196.

Singer, Charles. 1957. *A Short History of Anatomy and Physiology from the Greeks to Harvey*. New York: Dover.

Sinha, A. H. 1974. "How Passive are Passives?" In *Papers from the Regional Meeting of the Chicago Linguistic Society*, Chicago, 1974, pp. 631–642.

Sinsheimer, Robert L. 1968. Review of *The Double Helix* by James Watson. *Science and Engineering* (September 1968): 4, 6.

Small, Henry G. 1978. "Cited Documents as Concept Symbols." *Social Studies in Science* 8: 327–340.

Smith, Munroe. 1928. *The Development of European Law*. New York: Columbia University Press.

Sokal, R. R., and T. J. Crovello. 1984. "The Biological Species Concept: A Critical Evaluation." In *Conceptual Issues in Evolutionary Biology: An Anthology*, ed. E. Sober. Cambridge, Mass.: MIT Press, pp. 541–566.

Solem, A. 1978. "Cretaceous and Early Tertiary Camaenid Land Snails from Western North America (Mollusca: Pulmonata)." *Journal of Paleontology* 52: 581–589.

Solomon, Robert C. 1980. "Emotions and Choice." In *Explaining Emotions*, ed. Amélie Oksenberg Rorty. Berkeley: University of California Press, pp. 251–281.

Spector, Mark, Steven O'Neal, and Efraim Racker. 1980a. "Phosphorylation of the β Subunit of Na^+K^+-ATPase in Ehrlich Ascites Tumor by a Membrane-bound Protein Kinase." *Journal of Biological Chemistry* 255: 8370–8373.

———. 1980b. "Reconstitution of the Na^+K^+ Pump of Ehrich Ascites Tumor and Enhancement of Efficiency by Quercetin." *Journal of Biological Chemistry* 255: 5504–5507.

———. 1981. "Regulation of Phosphorylation of the β-Subunit of the Ehrich Ascites Tumor Na^+K^+-ATPase by a Protein Kinase Cascade." *Journal of Biological Chemistry* 256: 4219–4227.

Spector, Mark, Robert B. Pepinsky, Volker M. Vogt, and Efraim Racker. 1981. "A Mouse Homolog to the Avian Sarcoma Virus *src* Protein is a Member of a Protein Kinase Cascade." *Cell* 25: 9–21.

Sprat, Thomas. 1667. *History of the Royal-Society of London, For the Improving of Natural Knowledge*. London: J. Martyn and J. Allestry.

Stigler, George J. 1965 [1955]. "The Nature and Role of Originality in Scientific Progress." In *Essays in the History of Economics*. Chicago: University of Chicago Press, pp. 1–15.

Stimson, Dorothy. 1948. *Scientists and Amateurs: A History of The Royal Society*. New York: Henry Schuman.

Strawson, P. F. 1977 [1971]. *Logico-Linguistic Papers*. London: Methuen.

———. 1974. *Subject and Predicate in Logic and Grammar*. London: Methuen.

Style Manual for Biological Journals. 1964. 2nd ed. Washington, D.C.: American Institute of Biological Sciences.

Sulloway, Frank J. 1982. "Darwin's Conversion: The *Beagle* Voyage and Its Aftermath," *Journal of the History of Biology* 15: 325–396.

———. 1985. "Darwin's Early Intellectual Development: An Overview of the *Beagle* Voyage (1831–1836)." In *The Darwinian Heritage*, ed. David Kohn. Princeton, N.J.: Princeton University Press, pp. 121–154.

Swerdlow, Noel M. 1976. "Pseudodoxia Copernicana: or Enquiries into Very

Many Tenets and Commonly Presumed Truths, Mostly Concerning Spheres." *Archives Internationales d'Histoire des Sciences* 26: 108–158.

Swinburne, Richard. 1984. "Personal Identity: The Dualist Theory." In *Personal Identity*, a debate between Sydney Schoemaker and Richard Swinburne. Oxford: Basil Blackwell, pp. 1–66.

Thompson, John B. 1982. "Universal Pragmatics." In *Habermas: The Critical Debates*, ed. John B. Thompson and David Held. Cambridge, Mass.: MIT Press, pp. 116–133.

Tigar, Michael E., and Madeleine R. Levy. 1977. *Law and the Rise of Capitalism*. New York: Monthly Review Press.

Toulmin, S. 1977 [1972]. *Human Understanding: The Collective Use and Evolution of Concepts*. Princeton, N.J.: Princeton University Press.

Tufte, Edward R. 1983. *The Visual Display of Quantitative Information*. Chesire, Conn.: Graphics Press.

Turner, Edwin L. 1988. "Gravitational Lenses." *Scientific American* 259: 54–60.

Turner, Frank Miller. 1974. *Between Science and Religion: The Reaction to Scientific Naturalism in Late Victorian England*. New Haven: Yale University Press.

Turner, Victor. 1978 [1974]. *Dramas, Fields, and Metaphors: Symbolic Action in Human Society*. Ithaca, N.Y.: Cornell University Press.

———. 1981 [1968]. *The Drums of Affliction: A Study of Religious Processes among the Ndembu of Zambia*. Ithaca, N.Y.: Cornell University Press.

———. 1967. *The Forest of Symbols: Aspects of Ndembu Ritual*. Ithaca, N.Y.: Cornell University Press.

———. 1982. *From Ritual to Theatre: The Human Seriousness of Play*. New York: Performing Arts Journal Publications.

Vogt, V. M., R. B. Pepinsky, and E. Racker. 1981. "Src Protein and the Kinase Cascade." Letter, *Cell* 25: 827.

Waddington, Conrad H. 1968. "Riding High on a Spiral," review of *The Double Helix* by James Watson. *The Sunday Times* (London), May 25, 1968, p. 1.

Wallace, William A. 1959. *The Scientific Methodology of Theodoric of Freiberg: A Case Study of the Relationship between Science and Philosophy*. Fribourg: The University Press.

Waller, Robert H. 1979. "Four Aspects of Graphic Communication." *Instructional Science* 8: 213–222.

———. ed. 1979. *Processing of Visible Language*, vol. 1. New York: Plenum.

Ward, W. Dixon. 1967. Letter, *Physics Today* 20 (January 1967): 12.

Watson, James D. 1966. *The Double Helix: A Personal Account of the Discovery of the Structure of DNA*. New York: Atheneum.

Watson, J. D., and F. H. C. Crick. 1954. "The Complementary Structure of Deoxyribonucleic Acid." *Proceedings of the Royal Society* A, 223: 80–96.

———. 1953a. "Genetic Implications of the Structure of Deoxyribonucleic Acid." *Nature* 171: 964–967.

———. 1953b. "A Structure for Deoxyribose Nucleic Acid." *Nature* 171: 737–738.

———. 1953c. "Structure of DNA." *Cold Spring Harbor Symposia on Quantitative Biology* 18: 123–131.

Watson, James D., and John Tooze. 1981. *The DNA Story: A Documentary History of Gene Cloning.* San Francisco: W. H. Freeman.

Weigert, Andrew. 1970. "The Immoral Rhetoric of Scientific Sociology." *American Sociologist* 5: 111–119.

Weld, Charles Richard. 1858. *History of the Royal Society, with Memoirs of the Presidents. Compiled from Authentic Documents*, 2 vols. London: John W. Parker.

Wenzel, J. W. 1982. "On Fields of Argument as Propositional Systems." *Journal of the American Forensic Association* 18: 204–213.

Westfall, Richard S. 1984 [1980]. *Never at Rest: A Biography of Isaac Newton.* Cambridge: Cambridge University Press.

Westman, R. S. 1972. "The Comet and the Cosmos: Kepler, Maestlin and the Copernican Hypothesis." In *Colloquia Copernicana 1. Studia Copernica V*, Études sur l'audience de la Théorie Héliocentrique. Wroclaw [Breslau]: Polska Akademia Nauk, pp. 7–30.

———. 1975a. "Michael Maestlin's Adoption of the Copernican Theory." In *Colloquia Copernicana IV. Studia Copernicana XIV*, L'audience de la Théorie Héliocentrique Copernic et le Développement des Sciences Exactes et Sciences Humaines. Wroclaw [Breslau]: Polska Akademia Nauk, pp. 53–63.

———. 1975b. "The Wittenberg Interpretation of the Copernican Theory." In *The Nature of Scientific Discovery: A Symposium Commemorating the 500th Anniversary of the Birth of Nicolaus Copernicus*, ed. O. Gingerich. Washington, D.C.: Smithsonsian Institute Press, pp. 393–429.

Willard, Charles Arthur. 1983. *Argumentation and the Social Grounds of Knowledge.* University, Ala.: University of Alabama Press.

Williams, Joseph M. 1985. *Style*, 2nd ed. Glenview, Ill.: Scott, Foresman.

Wilson, Kenneth G. 1971. "Renormalization Group and Critical Phenomena. I. Renormalization Group and the Kadadoff Scaling Picture"; "Renormalization Group and Critical Phenomena. II. Phase-Space Cell Analysis of Critical Behavior." *Physical Review* B, 4 (November 1, 1971): 3174–3205.

Wimsatt, W. C. 1980. "Reductionist Research Strategies and Their Biases in the Units of Selection Controversy." In *Scientific Discovery: Case Studies*, ed. T. Nickles. Dordrecht: D. Reidel, pp. 213–259.

Wittgenstein, L. 1965 [1958]. *Preliminary Studies for the "Philosophical Investigations." Generally Known as the Blue and Brown Books.* New York: Harper and Row.

Woese, Carl R. 1967. *The Genetic Code: The Molecular Basis for Genetic Expression.* New York: Harper and Row.

Woolgar, Steve. 1981. "Discovery: Logic and Sequence in a Scientific Text." In *The Social Process of Scientific Investigation*, ed. Karin D. Knorr, Roger Krohn, and Richard Whitley. Dordrecht: D. Reidel, pp. 239–268.

Wright, Patricia. 1977. "Presenting Technical Information: A Survey of Research Findings." *Instructional Science* 6: 93–134.

Zarefsky, D. 1982. "Persistent Questions in the Theory of Argument Fields." *Journal of the American Forensic Association* 18: 191–203.

Ziman, J. M. 1968. *Public Knowledge: An Essay Concerning the Social Dimension of Science*. Cambridge: Cambridge University Press.

Zimmerman, David W. 1982. "Are Blind Reviews Really Blind?" *Canadian Sociology* 23: 46–48.

Zuckerman, Harriet. 1977. *Scientific Elite: Nobel Laureates in the United States*. New York: The Free Press.

Zuckerman, Harriet, and Robert K. Merton. 1973. "Institutionalized Patterns of Evaluation in Science." In Robert K. Merton, *The Sociology of Science: Theoretical and Empirical Investigations*, ed. Norman W. Storer. Chicago: University of Chicago Press, pp. 460–496.

Index

Social drama (*cont.*)
  and rhetoric, 181
  and symbolism, 185–186
  phases, 180–181; breach, 183–184;
    crisis, 185–188; redressive action,
    188–189; reintegration, 189–190
Sociology, aims contrasted with rhetor-
    ical analysis, 177–179
Speech act theory and science, 129–143,
    148. *See also* Ideal speech situation
Sprat, T.
  on scientific style, 17
  History, 165–168
  aims of science in, 182
Stasis theory. *See* Rhetorical terms
Stigler, G., on priority in science, 176–
    177
Strawson, P. F., and style in science, 70
Style. *See* Rhetorical terms: style
Swinburne, R., and personal identity,
    158
Synecdoche. *See* Rhetorical terms: style

Tables. *See* Visuals
Taxonomy. *See* Sciences
Topics. *See* Rhetorical terms: invention
Toulmin, S., Copernican revolution in,
    214n2
Tu quoque. *See* Rhetorical terms

Turner, V.
  role of naming in, 45
  and calculus priority dispute, 169–172
  social drama in, 180–181, 189–190
  communitas in, 190

Universal audience. *See* Perelman, C.

Visuals
  in taxonomy, 36–38, 45; in astronomy,
    106
  to increase presence in science, 63–64;
    and ontology in science, 74–80; and
    causality in science, 75–79

Watson, J.
  double helix paper, 28–29, 62–64;
    compared with *Double Helix*, 64–65
  Double Helix as an account of science,
    55–58; as a fairy tale, 58–62; as
    Freudian tale, 60–61; as theory, 64–
    65; compared with double helix
    paper, 64–65
  and recombinant DNA debate, 184,
    186, 187–188
Wittgenstein, L.
  on family resemblance, 34
  and objectivity in science, 50–51
Woolgar, S., on scientific papers, 85, 89,
    140